The
Charismatic
Personality

Dr Len Oakes

www.
AUSTRALIANACADEMIC**PRESS**
.com.au

First published in 2010
Australian Academic Press
32 Jeays Street
Bowen Hills Qld 4006
Australia
www.australianacademicpress.com.au

National Library of Australia cataloguing-in-publication entry:

Author:	Oakes, Len.
Title:	The charismatic personality / Len Oakes.
ISBN:	9781921513466 (pbk.)
Notes:	Bibliography.
Subjects:	Charisma (Personality trait)
	Leadership.
	Leadership--Psychological aspects.
	Leadership--Religious aspects.
	Political leadership.
Dewey Number:	303.34

To Noeline and Sefronia

• • • •

CONTENTS

• • • • •

continued over

CONTENTS *continued*

• • • • •

INTRODUCTION

· · · · ·

It is odd that so many use the word 'charisma' without under-standing its meaning. Perhaps they think that something so apparently magical and mysterious cannot be understood. Nevertheless, this book is about charisma; charismatic personalities, leaders, followers and movements, and the stance taken is that charisma is not magical or mysterious, it can be analysed, and that it tells us something profound about human nature. However, while it would be nice to synthesise extensive and wide-ranging research, unfortunately there has been so little rigorous investigation of charisma that such an effort would have only specialist appeal. Instead, this book aims to provide a general overview of the topic, and it draws from a range of disciplines from theology to psycho-analysis to provide a comprehensive account.

From historical and biblical sources and from sociological theory, an outline of the function of charismatic leaders and their move-ments is presented. From the biographies of charismatic personalities an analysis of their psychological natures and development is con-structed, and this is in turn filtered through psychoanalytic theory to probe their underlying motivations. From investigations into new religions and other social movements an understanding of the social-psychological dynamics of charisma is presented. From research into organisations and corporate leadership these perspectives are expanded, and from insider accounts by followers, an understanding of the meanings of charismatic relationships is provided. Such an overview may not satisfy purists from any of these disciplines, but hopefully it may inform a wider audience.

The argument advanced is that charismatic personalities are differ-ent from the general population in important ways, and that they are used by communities and societies to solve problems that reason and tra-dition have failed to answer. Hence, charismatic leadership is a creative, problem-solving strategy that is resorted to in extremis, and that the reason why it so often disappoints is partly because of the magnitude of the problems it is called upon to address. However, when it is successful it may be spectacularly so, and give birth to a new civilisation or religion.

The book is organised so that theory is balanced with case studies. Concepts from the scholarly literature are complemented with observations and ideas developed by the author, and examples are presented from historical and biographical material. The book is based on three major case studies: Winston Churchill, Sigmund Freud and Adolf Hitler. Later, several other figures are considered, including Mohandas Gandhi, Nelson Mandela, Germaine Greer, Girolamo Savonarola, Mao Tse-tung and Franklin Delano Roosevelt. These case studies and vignettes are one-dimensional, highlighting only the charismatic aspects, and are not intended to be comprehensive representations of these figures. The example of Churchill is returned to often because he epitomised so many of the central characteristics of charisma.

Part One outlines a theory of the psychology of charismatic personalities. The first chapter provides a brief overview of how the notion of charisma has been used in social and psychological theory, while the second chapter presents the case study of Churchill. Chapter Three introduces psychoanalytic insights from the work of Heinz Kohut, and this relates back to the case study of Churchill and forward to the following case study, that of Sigmund Freud presented at Chapter Four.

Chapter 5 extends the theoretical ambit to psychiatry and scientific psychology, and it especially addresses aspects of pathology that are relevant to charisma. This prepares the way for Chapter 6, which is a case study of an overtly pathological charismatic personality, Adolf Hitler. These particular case studies are chosen primarily because significant new material has recently emerged about them, specifically the biography of Winston Churchill's mother by Charles Higham (2006), the biography of Martha Freud by Katja Behling (2005), and the ground-breaking books on Hitler by Lothar Machtan (2002), David Lewis (2004) and Fritz Redlich (2000).[1]

Chapters 7 and 8 bring all this material together into a 'natural history' of the charismatic personality that describes the various life stages and major themes of charismatic development.

Part Two covers topics that derive from or expand these themes, and discusses the theory of Max Weber, who introduced the notion of charisma into sociology. Chapter 9 discusses the charismatic relationship of leader and follower. Chapter 10 considers charismatic movements and it is followed by chapters on charismatic women, the charismatic alliance and saintly charisma, each with appropriate case studies. A chapter on some remaining relevant aspects including charismatic communication, creativity, memory, social insight and the psychology of the will is presented to round out the subject. The book concludes with some evaluative observations.

Some additional explanatory notes are in order.

Currently, many scholars abjure the 'great man' theory of history (Worsley, 1970), most closely identified with historian Thomas Carlyle (2007). American historian Richard Pells put this fashion very well when he wrote:

> Ordinary readers (which means book buyers) are fascinated by the biographies of political and military leaders. They continue to believe charismatic personalities affect a country's destiny. Professional historians have long since abandoned that idea as a delusion. Instead, for the past 30 years, they have told us that the intricacies of social history are the key to explaining a nation's identity and development. (Pells, 2008)

Such a stance is, I believe, most sustainable among those who have never actually met a charismatic personality. To do so is to discover that they really are different; that they possess a mysterious 'presence' that others lack, along with several other extraordinary attributes. This is not to say that they dominate the world to such a degree that social history has no influence on society; of course it does. But this matter should not be reduced to an illusory either/or choice. Charismatic personalities are also products of their environments; but at certain critical points in history, the presence or absence of a particular person at the wheels of power may prove crucial. Despite the claims of Pells and others, in some disciplines such as political studies, the existence of strong personalities who are able to influence history has long been a taken-for-granted reality.

Indeed, it is indisputable. This book merely explains the central aspects of how such personalities arise.

Some readers may object to the use of psychoanalytic theory in a volume with a scientific allegiance. As Peter Watson has claimed, 'There is no inner self. Looking "in", we have found nothing — nothing stable anyway, nothing enduring, nothing we can all agree upon, nothing conclusive — because there is nothing to find' (Watson, 2005, p. 746). Such subjective nihilism is not restricted to nonpsychologists (Janda, 1996, p. 746), however it has no place herein. Charisma is one of several relatively common human experiences where the emergence of unconscious material into consciousness seems indisputable, rising seemingly out of nowhere to motivate action with a mysterious passion and conviction; comparable experiences include romantic love, creativity and mysticism. This is widely recognised by researchers in leadership studies, where even quantitative works are likely to begin with a psychoanalytic overview (Shamir et al., 2007).

This book grew out of my doctoral research and the subsequent book based on that research, titled *Prophetic Charisma: The Psychology of Revolutionary Religious Personalities* (Oakes, 1992, 1997). That work was a psychological study of twenty leaders of alternative and new religious movements, undertaken from 1989 to 1992. This book applies the ideas developed in that context more generally to charismatic personalities in the world of politics and history. Where the original study is mentioned herein, the brief title *Prophetic Charisma* is used.

The inclusive pronoun 'he' is used throughout to avoid confusion and in recognition of the fact that the great majority of charismatic personalities are indeed male. However there are exceptions, and when female charismatic personalities are considered this practice is reversed. When discussing followers, both pronouns are interchanged, and where grammar permits, 'they' and 'them' are used in singular usages.

Finally, I hope that this book will be of interest to both academic and nonacademic readers. For the latter I have taken out the insistent citation behaviour required by the former, and for the former I have preserved the references as notes at the end of the book. And, because jargon is the refuge of the insecure, it is used here as little as possible.

Endnotes

1   Continuing a long tradition of partial, polarised and biased biographies, these books have their shortcomings. For example, Behling's book on Martha Freud draws on the overly sympathetic biographies of (Sigmund) Freud by Earnest Jones and Peter Gay but does not even mention the more critical (and authoritative) works of Louis Breger (2000), Peter Webster (1996), Jeffrey Masson (1984) and others. That Behling knows of these works is shown by her citing the volume of letters from Freud to Fliess edited by Masson (1985), and also by her detailed discussion of the case of Anna O (a personal friend of Martha Freud). She makes no mention of the work of Henri Ellenberger (1970) who first recovered the details of the case of Anna O. She also perpetuates some myths about Freud such as that 'his ideas met with intense disapproval' in Vienna (Behling, 2005, p. 141), and makes no mention of Frank Sulloway's (1979) corrective volume. Despite such limitations, even highly biased books may sometimes present useful truths. For example, Jung Chang and Jon Halliday's biography *Mao: The Unknown Story* (2005), although excoriated by scholars, nevertheless did shed new light on why Mao was so keen to fight America in Korea (like Osama Bin Laden he believed that America had no stomach for war). Jung and Halliday also drew on interviews with Soviet officers who were in North Korea during the Korean War, and from this they thoroughly debunked the notion that America was conducting biological war against North Korea, an idea that had been extant in some circles. Hence their book is cited, albeit with caveats.

• • • • •

PART **ONE**

• • • •

Theory and Case Studies

CHAPTER ONE

Varieties of
Charismatic Experience

*The paradigm established by Weber and Durkheim, and re-stated
by psychological theory, claims, in fact, that society is based upon a
deeply evocative communion of self and other, a communion that
offers not reason, but lived vitality. Without this electrifying blurring of boundaries, life no longer has its savour, action is no longer
potent, the world becomes colourless and drab.*

Charles Lindholm

Surely no English word is more widely misunderstood than
'charisma'. The term is used to describe anyone who is momentarily popular or mentionable. On television and in the press it
conveys the same celebratory vagueness as phrases like 'well-known
social identity' and 'noted media personality' and other descriptors of
transients famous for being infamous. There are charismatic corporations such as Virgin, charismatic fashion houses such as Versace,
charismatic buildings like the Sydney Opera House; there is charismatic footwear such as Nike, conservationists speak of charismatic
animals like elephants and cheetahs, there is a gourmet dish called
'Crayfish Charisma', and even the frail Pope John Paul II in his later
years was described as charismatic. Indeed, the modern period itself
has been called the 'age of charisma' (Schweitzer, 1984).

The common thread underlying these usages seems to be that
someone or something is seen as impressive in some indeterminate
but larger-than-life way, as when the beauty queen on her date with

Frank Sinatra described him as having 'something I have never seen in my life ... a halo around his head of stars ... I can't snap out of it ... fascinating' (Hesse, 1970). But this popular usage is far from the original meaning.

That some people are just plain different, in a consistent way that we reliably describe as 'charismatic' seems indisputable; but to date there has been no comprehensive theory of such people, although charismatic leadership is recognised and valued by all cultures and countries (Bass & Riggio, 2006, p. 16; Dorfman, Hanges, & Brodbeck, 2004). Some argue that, despite appearances, there is no charismatic personality 'type' (Conger & Kanungo, 1988; Strozier & Offer, 1985), although in the past, similar consistent behavioural patterns have been considered sufficient to identify such broad and variable personality types as extroverts and neurotics, and to guide study of them. In fact, there has been at least one attempt made to describe and quantify the main psychological dimensions of charisma (Burke & Brinkerhoff, 1981). This book proposes that there exists a psychological pattern that may be described as the charismatic personality, and it proposes a theory that may go some way towards explaining the natures of such persons.

Throughout history, and recapitulated in recent scholarship, certain recurring themes concerning charisma have been debated. The first concerns the notion of large versus small charisma. We all recognise personalities such as Winston Churchill and Franklin Roosevelt as charismatic, but are such individuals really so different from the rest of us, or might we all have some seed of the charismatic mentality? This is one of the questions underlying what little psychological research has been done into charisma (Burke & Brinkerhoff, 1981); it was an assumption made by St Paul when he spoke of the different 'gifts of the spirit' given by God to each of us (more on this later).

Then there is 'good' versus 'evil' charisma. In a recent book on transformational leadership, a term often used interchangeably with charisma, the authors use the term 'psuedotransformational' leadership to designate charismatic leadership that is not morally uplifting (Bass & Riggio, 2006, pp. 12–16). Unfortunately, this appears to overlook testaments such as Joseph Goebbel's diaries, to the effect that many Germans during the Nazi period experienced Adolf Hitler as very morally uplifting (Heiber, 1962).

Another recurring debate concerns whether the charismatic leader is really charismatic because of their own self and attributes, or because they have been 'socially constructed' by others to fill a leadership role. Certainly one can observe followers doing this; sometimes, even in the face of leadership failure, they will interpret events in such a way as to 'save' the leader (from himself) and thus perpetuate their own agendas. But this turns out to be one of those chicken-and-egg conundrums that bedevil social science and which ultimately has no answer, for both individual and group processes are necessary for the creation of a charismatic leader. The literature on 'follower-centered theories of leadership' covers this territory, although there is a tendency in such writing to think of leadership as merely a role, rather than to think of leaders as individuals (Shamir, Pillai, Bligh, & Uhl-Bien, 2006). At its most radical, this approach sees leaders as irrelevant and interchangeable (Shamir et al., 2006), a position that is clearly untenable. It also diminishes the leader's responsibility for his actions.

Then there are charismatic leaders who are not really charismatic at all; the sickly child whose feverish dream resonates with the needs of a tribe in crisis and who is raised by them to leadership, more on the basis of superstitious hope than on the evidence of any leadership skills. As well, there are charismatic non-leaders — Zorba the Greek being a prime candidate from literature — and charismatic personalities who never actually achieve significant leadership, perhaps because they lack key skills such as management ability, or perhaps because events bypass them and their message. There are both male and female charismatic personalities, and conservative and radical charismatic leaders; and while some are highly confrontational — Churchill, for example — others are not — typically Father Divine, founder of the Peace Mission. Harry Truman seemingly only developed charisma after a long and successful career as a non-charismatic politician, yet he achieved much, including the Berlin airlift, the Marshall Plan, initiating NATO, and devising the Truman Doctrine to defend Greece (Bass & Riggio, 2006, p. 82). And there are charismatic failures; not failures because their vision or their charisma fails, but for other reasons. The US Civil War General George B. McClellan was loved and idealised by the troops in his command, and was very effective in training and organising his army. But when it came to battle he was almost completely ineffective, despite having superior forces. He might have shortened the war by two years had he engaged with General Lee, but his style of leadership was only effective in peace (Bass & Riggio, 2006).

Finally, there is no single behaviour that the charismatic personality displays all the time. Like the rest of us, most of their behaviour is situational; that is, they do the same things at the movies or when dining out or attending church that we all do, so it is impossible to identify a definitive trait or behaviour that unerringly demarcates them from others. Hence, it is sometimes difficult to distinguish between the charismatic personality and the merely strong personality, who may well also be colourful and dramatic, gritty and determined. In fact, there are reliable ways of distinguishing between the two — for example, that the charismatic personality lives through other people, whereas the merely strong personality does not — but such traits are not immediately apparent or easily measurable.

In essence, charisma involves the idea of a divinely inspired power or talent, and this notion is as old as mankind. The oldest surviving work of fiction, the *Epic of Gilgamesh*, tells of a warrior-king, part god and part man, who quests for the secret of eternal life. He has many adventures in the lands of the gods and even attains that which he seeks, only for it to turn out to be an illusion. He returns home convinced of the futility of his quest, and aware that 'the central fact of my life is my death'. Despite his divine gift and his numerous successes, in the end he dies a failure (Heidel, 1968; Kopp, 1972).

Our word 'charisma' comes from ancient Greece, where Charis, the goddess of grace, personified beauty, purity and altruism. The Greek term is *charizesthai* and it means favour or gift of divine origin. Curiously, the Greeks do not seem to have associated this with the kind of demagogic and irrational leadership of which Plato wrote in his Gorgias, although they were well aware of the rhapsodic

'Dionysian' aspect of life; Plato himself was a member of the Elysian mystery cult. For Aristotle the megalopsychos was the great man who dared to live alone in secret worship of his own soul. The Romans called the hero's charismatic power *facilitas,* and they too believed that it came from the gods.

Later usages derive from St Paul, who also saw charisma as a gift from God, although he had a much less heroic account of it:

> To one there is given through the spirit the message of wisdom, to another the message of knowledge by means of the same spirit, to another faith by the same spirit, to another gifts of healing by that one spirit, to another miraculous powers, to another prophecy.
>
> (1 Corinthians, 12:8–10)

Because the Christian use of the word has become so widespread, even being used to designate the variant of Pentecostal religion, it is instructive to examine what Paul originally intended. Three things are consistent throughout his writings. First, everyone has some gift, great or small ('Every one has his proper gift from God ...' [1 Corinthians, 7:7]). Second, there is much diversity of such gifts and, in addition to wisdom, knowledge, healing, miraculous powers and prophecy, Paul also lists service, teaching, encouraging, giving to others, leadership, being merciful, administration, and speaking in tongues as divinely bestowed gifts — and there is no suggestion that this is an exhaustive list. Third, all such gifts are intended to be subordinate to 'the common good' (1 Corinthians, 12:7).

In this, Paul foreshadows our modern usage in that he lists a number of talents that may be gifts from God and he allows for the possibility of there being many more such gifts. These may not necessarily be associated with leadership, although he recognises that anyone with a prodigious talent may become a leader of sorts, be they male or female, old or young, healthy or sick. This is why he adds the proviso that such gifts must be used for the common good rather than in the service of the bearer's ego, for in Paul's thinking this is what God created these gifts for.

It is hard to fault Paul's formulation because clearly we are not all created equally (leaving aside the issue of a creator). Many people, from mathematicians to Olympic gymnasts, possess extraordinary talents that seem quite uncanny, and often such people do become leaders, if only among the community they grew up in, although perhaps equally often they do not. Paul also allows that just as there are great gifts, so too there are more modest gifts, yet these also come from God and thus are in a lesser way also charismatic.

A good example of such a modest charismatic gift might be the man who never does anything extraordinary yet who manages to survive well. In difficult times anyone who has good work, a sound marriage and a happy healthy family will be looked to for leadership, be they merely the kindly family doctor, or a wise and compassionate grandparent. Another example might be the artist or master crafts-man whose work carries a special aura of transcendence; we have all experienced feelings of awe and wonder when examining an excep-tional piece of art or craftsmanship. It seems that when someone is doing what they love, that which is their life's work and that they have invested their ideals in, observers respond with feelings of love and awe, and the worker is perceived as being blessed by God, or at least embodying something extraordinary.

Sometimes we ourselves experience moments when others look to us in appreciation of our skills or knowledge as if these denote something great. We are likely to feel embarrassed at such times and may mutter, 'It was nothing'. These moments are probably carry-overs from childhood when the little boy declared that he intended to grow up to be 'Just like dad', and the little girl said that when she grew up she would have two babies, 'Just like mum'.

Such quasi-charismatic sentiments — feelings of love and awe — are a normal reaction to apparently extraordinary people, including parents. It seems that we are primed by nature to detect authority and substantiveness in others, for that is how we learn and progress, although our judgment may be mistaken at times. When a good actor pretends to have such attributes, or when the behaviours of a dangerous misfit coincide with what we expect wisdom and authority to look like, we might find ourselves giving our trust and faith to someone unworthy, with disastrous consequences. Still, we must recognise that even such mistaken idealisations are variations on basically normal, healthy and realistic themes, including the need for loving parents, the quest for guides and supports, the search for wise and virtuous leaders, and the hope of transcendent meaning, all of which seem to be implicit in our make-up.

The most primitive form of charisma occurs in shamanism. This is the religion of the small tribal unit and the witch doctor. The shaman is 'one who is excited, moved, raised ... the master of the techniques of ecstasy' as Mircea Eliade (1964, p. 4) put it. Typically he (or she, for among the !Kung of the Kalahari, fully 10 per cent of women become shamans; Lindholm, 1990, p. 158), is identified early as one with a 'shadowed heart'. The shaman is not psychotic but is disturbed in some way, the 'disease of God' as the Koreans call it. He or she shows peculiar behaviours from birth and experiences spirit possession, trance and epileptic seizures while still young. Such a youth is apprenticed to a senior shaman who trains him or her in occult practices. After hearing a call from a god or a spirit the trainee withdraws into the desert or the woods to meditate in solitude, often undergoing some kind of spiritual test such as a journey to the underworld. This culminates in a spiritual rebirth from which the shaman emerges with an inner strength and uncanny sensitivity, emotional intensity and detachment; in sum — charisma. Transformed, the graduate shaman returns to the tribe to claim his or her place as tribal witchdoctor.

Thus the shaman is a 'wounded healer' who has conquered a sickness and learned to use it for the benefit of others. He or she is able to explore sacred realms and to mediate with the spirit world on behalf of the tribe. Allied with this are the skills of psychopharmacology, healing and the mastery of trance states. These enable the shaman to preside over the ceremonies, ritual functions and crises of the tribe.

The shaman is unpredictable and fearless, holding office by virtue of his personal spiritual attainment and psychological voltage, and is credited by the tribe with mysterious and dangerous supernatural powers. The shaman's peculiar disturbance and training enables him to see through the illusions of his community and to diagnose its ills and prescribe social cures for its members. Anthropologist Weston

La Barre has described 'the eerily supernatural omniscience and compelling power of charisma streaming from the shaman like irresistible mana', and said that it comes from an ability to discern the unconscious wish-fantasies of the group, adding that the shaman 'is so unerringly right because he so pinpoints these wishes' (La Barre, 1980, p. 275). It is this power that earns the shaman his place, for he is feared rather than loved.

Another charismatic variant found in more complex societies is the prophet. In the Old Testament the prophet arose in a time of crisis, usually from among the people rather than the priesthood or the aristocracy, and he called the faithful back to the old ways by preaching a warning from God. In other societies prophets are also associated with crises, and they usually propose a solution that challenges the traditional and legal authorities. Further, once a tribe or society has incorporated the Christian notion of the prophet into its beliefs, typically by taking up Christianity through missionary activity, it becomes able to generate new prophets in the image of Old Testament prophets; there are dozens of such figures in contemporary Africa. The social niche of the prophet even enables the tribe to create prophets where there are none, temporarily elevating a 'false prophet' to the status of charismatic saviour, as in cargo cults. In more modern times, the niche of the prophet has enabled religious entrepreneurs from Asia to evangelise the West with Eastern religions.

Although charismatic leadership can occasionally arise in the absence of stress (Bass & Riggio, 2006), usually people only turn to charismatic leaders when facing a problem that they have been unable to solve by traditional or rational means. Nations usually only turn to them when the problem they are facing is especially acute, sometimes being prepared to risk grave danger by turning to an extremely unstable charismatic figure such as Adolf Hitler as their saviour. There is even a particular kind of social organisation — the charismatic movement — that arises around charismatic leaders, involving intense idealisation, magical thinking and radical surrender, and instilling faith in a better future for all who join (Bass & Riggio, 2006, p. 39). So compelling is such energised hopefulness that, in a crisis, the delirious dream of the epileptic child may take on totemic dimensions; the child being taken up by the group and worshipped as the saviour until the inevitable disappointment occurs, but this is not charisma in the sense intended in this book.

The great German sociologist Max Weber studied charisma and introduced the concept into sociology. Weber defined charisma as 'a certain quality of an individual personality by virtue of which he is considered extraordinary and treated as endowed with supernatural, superhuman, or at least specifically exceptional powers … [that] are regarded as of divine origin' (Weber, 1968b, pp. 241–242). This definition created a conundrum for social scientists because it seemed contradictory; on the one hand, it specifically defined charisma as a 'quality of an individual personality', but on the other hand it invoked the criterion of the social group as to whether or not an individual is 'considered extraordinary'. Purists demand that charisma be defined as either a psychological trait or a construction by the group; would that things were always so neat.

In contemporary society we all know of natural leaders, those gifted men and women who seem to have the golden touch. They are confident, charming and popular, energetic and smart. They come to head countries, corporations, and religious and other groups. They are inspirational and seem to promise much, and they attract others whose aspiring hopes need a focus or a support. Yet despite their great talent, sometimes such figures do not have a very good record of sustained success. If they attain high office they may sabotage their own best efforts, as Bill Clinton did. If they head large corporations they may overreach and self-destruct through rash decision-making, as Thakshin Shinawatra of Thailand recently did. If they lead a religious group there may be scandal or worse, as with Bhagwan Shree Rajneesh. Sometimes the followers and associates of such personalities emerge scarred and disillusioned from their associations.

As well, when we examine the personal lives of such individuals we find that often there are major inconsistencies, such as antisocial behaviour; emotional problems, such as alcoholism or depression; or there may be infidelity, or a reign of terror in the home. It is as if the leader has one benign and successful face for their followers and another discordant, contradictory face for their private lives. They almost never have close friends. They tend to polarise people, evoking love and devotion among their followers, while generating hatred among critics and defectors.

There is a paradox here that deserves explanation. A particular and recurring kind of personality — easily recognisable by such key traits as immense self-belief, enormous energy and great charm — who seems to be a natural leader and who rises to a position of dominance, turns out to be unstable and may end up hurting others or self-destructing, despite their expertise. It is as if charisma involves some kind of psychological dysfunction.

This is the charismatic personality, a combination of genius and madness, yet possessing an easy mastery for manipulating the hopes and fears of others. The charismatic personality is drawn to people seeking leaders and solutions to problems, a natural enough impulse on their part, but one that is vulnerable to exploitation by the charismatic personality who 'knows' he has the answer to their problems.[1] These may be relatively minor problems, such as running a business or a church, coaching a sports team, or just providing entertainment or selling door-to-door; or there may be huge problems such as leading a nation through crisis or war. The charismatic leader may actually solve the problems, but there is always a risk because he has problems of his own, and sometimes he is not in great psychological shape.

Hence, many people have wondered if perhaps there is some mental illness that makes some disturbed people more, rather than less, likely to succeed in society. They wonder this because sometimes their leaders seem to behave so irrationally. This corresponds with the suspicions of many that modern life is in some way 'sick'. To discover that indeed, there is such an illness, and that charismatic personalities embody the psychodrama of the masses, turns a curious scepticism into something much more worrying.

To summarise, charisma means much more than mere popularity or infamy. Charisma has existed at all times and places in various forms. It exists wherever there is a human relationship that feels extraordinary or even divine, and it is sharpest where leadership in crisis is involved. Its psychological origins lie in infancy and are based on a natural tendency to respond with love and awe to what is extraordinary, strong and benign. At the simplest level this may involve a child's idealising of a parent. At other times, such feelings may be inspired by quite ordinary people doing apparently extraordinary things, to which we again respond with awe and love. But sometimes a particular kind of personality is involved, the kind of person who is attracted to problems and who feels entitled to lead others. We may not ourselves respond with feelings of awe and love to them, but we recognise that others do; to them they are charismatic.

Thus, there are great and small variants of charisma, from the revolutionary call of the Messiah, to the winning ways of the new boss, to the sublime substantiveness of the craftsman; from the shaman to the religious reformer, and from the guru to the demagogue. What underlies all of these roles and relationships is an experience of the spiritual; an encounter with something that is deemed 'ultimate' in some way and that transforms ordinary life, leading to joy and freedom from that which is stale, routine and outmoded, although it may also run the risk of disappointment, or worse.

Endnote

1   Britton (1998) has coined the term 'epistemic narcissism' to denote an unshakeable belief in the rightness of one's own ideas, and added that it is the mark of a creative and assertive self, thus indicating that at least under some circumstances it may be quite normal and healthy.

CHAPTER TWO

• • • •

Winston Churchill: An Exemplar

First and foremost he was incalculable. He ran true to no form. There lurked in every thought and word the ambush of the unexpected. I felt also that the impact of life, ideas, and even words upon his mind was not only vivid and immediate but also direct. Between him and them there was no shock absorber of vicarious thought or precedent gleaned either from books or other minds. His relationship with all experience was first-hand.

Violet Bonham Carter (1965)

The life of Winston Churchill shows the charismatic personality at its best. Despite a damaging childhood he went on to play a spectacular historic role that confirmed his boyhood dreams of heroism and greatness. In such extraordinary success there can be healing for the damaged soul and, in Churchill's case, this seems to have occurred.

Churchill was the grandson of a duke, and in the hierarchy of the British aristocracy a duke is beneath only the royal family. Thus Churchill, although he inherited no title, was raised within the uppermost social class. This must be kept in mind because at least some of his ambition and sense of entitlement, and also his sense of duty, were quite normal for one so born to rule, or at least they were for many. Unfortunately, this was also a time of utter decadence and corruption within the ruling elite of Britain, or so it has to appear to anyone reading with a modern sensibility. Routine sexual infidelity, colossal fraud in government finances, military adventurism for private gain, and shameless hypocrisy in political and public life were all staples of Winston's social background, so much so that — if only for nothing else — he could be applauded for being unconventionally honest and decent in most of his dealings with others. He could easily have turned out quite miserable or nasty, but instead he accepted that sense of extended responsibility that the best of the privileged classes espouse: to whom much is given, from whom much is expected,[1] and he sought his destiny within conventional morality.

Churchill's mother Jennie seems to have been hugely influential in this, and she was herself one of the great characters of history. Her father was a corrupt and decadent American robber-baron whose customary diet was oysters and champagne, and Jennie received a cultured education, becoming an excellent pianist and finishing her schooling in Paris. She was beautiful, energetic, ambitious, highly intelligent and wilful, but, as Charles Higham, author of a brilliant recent biography of her wrote, 'there was something of the dominatrix about her' (Higham, 2006, p. 36). Steered by her father towards the British aristocracy she married the untitled wastrel Randolph Churchill after they conceived Winston out of wedlock. She later coached and led Randolph into a successful career in politics, writing his speeches for him and organising his campaigns. She took an early interest in criminal justice, and prison reform became a lifelong passion for her; later she influenced Winston in penal reform when he was a minister in government.

Jennie and Randolph's marriage soon became an 'arrangement' in which each went their separate ways. She was highly promiscuous, knowledgeable in 'every French vice', and was reputed to have had 200 lovers, although the real figure was almost certainly lower. She was probably mistress of the Prince of Wales, later King Edward VII, for a while. Another of her admirers, the Earl of Falmouth, after she resisted him, raped her on her drawing-room floor, but she forgave him and treated him warmly for the rest of her life.

Jennie performed several major good works in her life, and engaged in several misadventures. During the Boer War she funded and operated a hospital ship to care for wounded soldiers; later during World War I she funded and operated another medical facility for soldiers. She started a journal (it soon failed), wrote and produced a play (it too failed), launched a large-scale cultural festival (it also failed), espoused animal rights, and was swindled by a conman out of a large amount of money, after which she was maintained by the rich banker Natty Rothschild (partly in return for political favours). She published several books (one of which purported to be autobiographical but was full of lies), and was the inspiration for several novels. She secured important positions for young Winston and later organised his political campaigns. After Randolph's early death she married a much younger man who suffered from manic depression (she wrote letters of encouragement to him while Randolph lay dying in the next room), whom she scandalously divorced in 1913. She then married again, this time to an even younger man 24 years her junior.

The great mark against her was the annexation of Upper Burma in 1885. This was ruthless greed at its most transparent. Upper Burma was rich in teak and rubies and Queen Victoria and Natty Rothschild had their eyes on these, so Jennie — through Randolph as Secretary for India — orchestrated the annexation for which she received from Victoria the Order of the Crown of India. It has to be admitted that North Burma was at that time misruled by the corrupt King Thibaw, and that British rule had to be better for the ordinary Burmese, but these were not considerations that were foremost in her mind. She also espoused the Irish cause and happily offended the Tory establishment, which shunned and forgave her several times.

Jennie gambled away the money for her sons' educations and subsequently sued them to prevent them seizing assets to which they were legally entitled; this episode was hushed up by the courts. It seems she was always loyal in her affections, but she played hardball with everyone. Perhaps her spirit is best conveyed by her behaviour when, at age 61, after breaking her ankle and then having gangrene set in, she was told that she would have to have her leg amputated. Although those with her became hysterical, she coolly urged the surgeon to make sure he cut high enough, then quipped that in future she would just have to put her best foot forward. She died later that year.

With such an extraordinary mother, Winston would have had access to an exceedingly rich and potent conceptual repertoire to draw upon. Curiously, although he worshipped his mother, he came to yearn more for love from his father, the weaker of his parents. Randolph would have been nothing without Jennie, and she made him into one of the most exciting and progressive politicians of his time, although despite her tutelage he remained something of a loose cannon throughout his career. He established a radical cell within the Tory party and championed justice for Ireland. He uncovered tremendous graft and corruption in military funding and took a principled stance against it, eventually resigning in protest; but his resignation was done independently of Jennie, who would certainly have proscribed such an action. He narrowly avoided several scandals, made some major political blunders that he managed to wriggle out of, usually with Jennie's help, and in 1887 undertook his most ambiguous adventure (with Jennie), a diplomatic junket to Russia and Germany that, on the one hand was so irresponsible that it might have led to war, yet which actually helped secure peace with Russia for many years (Higham, 2006). Whether he (or she) understood the full significance of his actions is not known.

Winston's childhood had the potential to stunt him forever. In his celebrated biography *Churchill*, Roy Jenkins describes Winston as utterly neglected by his parents, referring to his 'semi-relationship' with his mother and 'non-relationship' with his father (Jenkins, 2001). From infancy Winston was raised by his devoted nanny Elizabeth Everest (convention designated her 'Mrs' even though she was unmarried), who was childless and poor and who seems to have poured all her affection into him and his younger brother. He only saw his parents in the evenings for a short time, after having been appropriately scrubbed and dressed. On many such evenings his father was not present and often nor was his mother. He and his brother would sit together for dinner with Everest and a tutor for company while their parents partied elsewhere.

As a small child longing for parental contact, Winston created a sustaining myth about his parents, choosing to interpret their frequent absences and indifference as being due to their being engaged in important work. He convinced himself that his father was a great man and that his mother was as important as she was beautiful. In later life he recalled his mother as 'a fairy princess, a radiant being possessed of limitless riches and power. She shone for me like the evening star. I loved her dearly but at a distance' (Carter, 1965, p. 25).

Winston created a magnificent dream for himself to ease his loneliness and to give him hope for some day conquering his parents' hearts; he dreamt of being a great hero. He had over 1,500 tin soldiers, and he spent hours mustering them into armies and pitting them against each other. He survived his parental neglect by living this fantasy world while yearning for the time when he could become the great man he longed to be, and thus earn his parents' attention and love.

The psychohistorian Jerrold Post has dubbed such self-bolstering strategies 'compensatory grandiosity', and they crop up in the lives of many — perhaps most — charismatic figures, usually quite early (Post, 2005). The logic of such a strategy is that Churchill probably guessed his parents' indifference to him but found it too painful to accept, so he denied it by inventing his myth that they really did love him but were engaged on important business that took them away from him. This gave a positive interpretation to their neglect and bolstered his self-esteem. There was some comfort in being the beloved child of ones so great and important, but it did not entirely dispel the painful truth. However, he also sensed that as their child he too was potentially great, despite his deprivation. So he constructed a life script in which some day, through his own greatness, he would receive the love he craved. If he could just become great he could gain entry to their world and thus 'prove' that they did indeed love him, despite their ongoing neglect.

This is 'magical thinking', an infantile fantasy process in which, if Winston could just wish hard enough, he might redeem the painful past/present in some near/far paradise wherein he and his loved ones always were, now are, and soon will be. By succeeding in the future, by becoming great and gaining his parents' love, he would refute the painful truth that they did not really care for him; he would prove his fantasy explanation of their indifference to be true, despite cold hard reality. Now he felt the beckoning of destiny, and every further act of parental neglect only crystallised deeper in his mind the proof of his calling. He could be sure of his parents' love only by becoming great.

It is also likely that this myth originated with the benign Mrs Everest having to explain to young Winston, yet again, why his parents chose to be away from him. We can almost hear her saying, 'They have very important work to do, my boy, as you will have too, one day'. Thus we might perhaps credit her with much of his future political drive and accomplishment. Certainly, she seems to have been his humanising saviour. He experienced her as wonderfully warm and loving, waxing lyrical about her in his later autobiography *My Early Life* (1930). She was his 'comforter, his strength and stay, his one source of unfailing human understanding ... the fireside at which he dried his tears and warmed his heart. She was the night-light by his bed. She was security' (Carter, 1965, pp. 25–26). It seems that Everest was genuinely able to derive great satisfaction from self-less service to others, especially children, with no residue of envy, rancour or self-pity, and without expectation of great reward. She could be indulgent and meddlesome (Higham twice refers to her as a 'mollycoddler'), but she stayed on with the family long past Winston's childhood; he refused to give her up even as an adult, and he kept a photograph of her on his bedroom wall till the day he died. Perhaps he felt a kindred spirit with her; someone who, like him, had been dealt a rum hand in life but who soldiered on anyway doing what best she could, asking no favours and bearing no grudges. Perhaps the example of her generosity of spirit defused his rage at his parental neglect and channelled his creative talents into a prosocial direction.

At age seven, Winston was increasingly a handful, a young egomaniac, loud and offensive, and it is at this point that descriptions of his childhood polarise. Jenkins described Jennie and Randolph as being cold, distant and unavailable towards Winston, while Higham said they 'were very good to him' (Higham, 2006, p. 78). Higham reveals his thinking in the passage: 'Life for him was bliss; although the Churchills, with House of Commons commitments, were not often at Blenheim [where Winston lived], no child was more thoroughly spoiled' (Higham, 2006, p. 79). In fact, it may have been exactly that absence and spoiling, plus Jennie's dominatrix demeanour and Randolph's personal limitations — as Higham has described — that produced Winston's disturbance, for disturbed he certainly was (Higham, 2006, p. 36). However, he did have his comforts, including his younger brother Jack, wealth, spacious grounds to play in, and an adoring aunt in the Duchess of Marlborough, who gave him affection, lessons and indulged him with treats.

Then Winston was sent to St George's Boarding School at Ascot, described by Jenkins as one of the most brutal in a brutal era. Jennie had herself been raised under the harsh discipline then believed to put steel into the ruling class, a regime administered through the private boarding-school system and intended to instil a hypermasculine, win-at-all-costs attitude, along with genuine scholarship and physical prowess as required for service to the Empire. This was a culture of aristocratic savagery at its worst, and under it he seems to have become quite a repellent creature, a dishonest, troublesome, endlessly complaining and manipulative child who borrowed money from other boys and would not pay it back, became a bully, learned little, and was 'rather greedy at meals' (Higham, 2006, p. 93).

From St George's, Winston wrote piteous letters home begging his parents to visit him and promising to try harder at his school work if they did. They usually did not, and in time he became quite rebellious and was thrashed often. He floundered at the bottom of his classes and developed a lifelong speech impediment, probably due to the anxiety he suffered at this time. His state of mind and academic performance only improved after he attended Harrow, where he met teachers who recognised his talents for history and language. Unfortunately, it was at this point that his father gave up on him and seems to have actually come to hate him.

Winston developed a phenomenal memory. While still a junior boy at Harrow he won a prize open to the entire school by memorising and reciting 1,200 lines from Macaulay's *Lays of Ancient Rome*. After this he warmed to his studies and began to succeed. It was a turning point that gave him a skill he could develop and use to get ahead, as well as earning him the respect of sympathetic teachers who were to function as mentors for him.

A remarkable and very telling adventure happened when Winston was in his last year of school. He was playing with his younger brother and a cousin at home and they were chasing him and had trapped him in the middle of a footbridge over a deep ravine. He knew that there could be no escape, or could there?

> In a flash there came upon me a great project. The chine which the bridge spanned was full of young fir trees. Their slender tops reached to the level of the footway. 'Would it not' I asked myself, 'be possible to leap on to one of them and slip down the pole-like stem, breaking off each tier of branches as one descended, until the fall was broken?' I looked at it. I computed it. I meditated. Meanwhile I climbed over the balustrade. My young pursuers stood wonderstruck at either end of the bridge. To plunge or not to plunge, that was the question! In a second I had plunged, throwing out my arms to embrace the summit of the fir tree. The argument was correct, the data was absolutely wrong. It was three days before I regained consciousness and more than three months before I crawled from my bed. The measured fall was twenty-nine feet on to hard ground. (Churchill, 1930, p. 30)

Even after finishing school it still took him three attempts before barely scraping through the army entrance exams. Unfortunately, his father Randolph suffered from Raynaud's disease and died from a brain tumour in 1895 at the age of 46 (Higham, 2006); soon after Winston was accepted into the military, so he never got to show off his brilliance to his father, and he was left with a morbid fear that his own life would be similarly short-lived. He later wrote: 'All my dreams of a comradeship with him, of entering parliament at his side and in his support, were ended. There remained for me only to pursue his aims and vindicate his memory' (Carter, 1965, p. 23).

This was Winston's perception at the time, or at least, it is his report of that perception some years later, but he actually lost both his father and Mrs Everest within a year — her employment was terminated — and it is likely that his grief for one is colouring his image of the other. The reality is that his father had written to him expressing disapproval of his enrolment in the military academy at

Sandhurst (it was a cavalry unit and he preferred the infantry), and predicting that Winston was destined to become 'a social wastrel, one of the hundreds of public school failures … you will degenerate into a shabby, unhappy and futile existence' (Higham, 2006, p. 146). Apparently this letter burned into Winston for the rest of his life. He had also been mistakenly informed by the family doctor that his father had contracted syphilis, and for the rest of his life he probably believed that his father had died from this. His mother also seems to have become embittered towards him at his point, expressing her complaints in letters to him and reinforcing these with her absence. He was now on his own.

Churchill studied hard at Sandhurst and eventually graduated 8th in his class of 150 officers. He joined the colonial army in India and sought the high road to fame with unabashed confidence, living a *Boy's Own* adventure of daring and fortune, while in his spare time writing a novel and a history of his experiences, and studying parliamentary debates to prepare him to follow in his father's footsteps as a politician some day. He saw action in Cuba, then became embroiled in a military scandal that saw him and some officer friends castigated in the press for race fixing, bullying and the ruination of two young men's careers.

However, he soon distinguished himself with his deeds of derring-do as a cavalryman undertaking fierce and apparently fearless sabre charges in Egypt, which he also wrote about in his massive book *The River War*. Later, in the South African War, he made a famous escape from the Boers, which he also wrote about at length in his now burgeoning secondary career as a journalist and war correspondent. (Much later his account of his escape was qualified by others who had been imprisoned with him, and who claimed that he had sabotaged their intended mass-breakout by his impulsive behaviour.) These despatches made him world famous as a celebrity speaker, and a writer by age 25, although his speech impediment ensured that he had to practise for hours, sometimes days or even weeks, before delivering a speech.

Because his father had been a politician, Churchill developed the notion that politics was the highest calling. He seems to have thought that excelling in his father's career would thus win him the affection and respect he had never received. Hence, after his discharge from the army, in an act of filial worship he entered politics, where he slavishly imitated his dead father's attitudes and behaviour, at least in his early years. Perhaps he even carried on an inner dialogue with his father's shadow. He wrote a two-volume hagiography of his father in an effort to prove to the world — or more likely to himself — what a great man his father had been. His mood in these years has been described by biographer Jenkins: 'His impatience, his self-centredness and his conviction that he had to pursue fame on every day of what he believed would be his short life all combined to give him a feeling of almost divine right to immediate freedom' (Jenkins, 2001, p. 56).

Churchill's maturity illustrates many of the complexities of charismatic personalities. Although charming, witty and benign he was not particularly likeable. His sense of self-importance was legendary; Lady Lytton said of him, 'The first time you meet Winston you see all his faults, and the rest of your life you spend discovering his virtues'. The triad of traits that are most consistently encountered in charismatic personalities — an inability to have equal friendships, the refusal to accept criticism or blame, and the need to be right and in control — were all well developed in him. He also talked too much and too loudly, had an 'uncontrollable desire for notoriety', and was 'devoured by vanity'. But he was also a supremely skilled arguer, almost impossible to defeat in any personal encounter.

By today's standards Churchill would probably be regarded as an alcoholic because he drank all day long, although he was a sipper, not a guzzler. His sense of entitlement was phenomenal, Jenkins remarking that 'if he had wanted a music lesson it would have been Sir Edward Elgar who would have been sent for, or if a little nursing attention had been required Florence Nightingale would have come out of retirement'. But his sheer energy was boundless and infectious, and in later years when facing the pressures of leadership in war, he was able to control his more extreme impulses (with assistance from his wife). In this early period he lived by phrase-making, but his parliamentary speeches were not consistently successful and he delivered many a clanger. He even traumatically forgot his lines once early in his career, a humiliation he took great pains to ensure never happened again. He was occasionally the butt of ridicule from both sides of parliament.

Churchill's marriage bears all of the hallmarks of a typical charismatic marriage in that he and Clementine seem not to have been very close and she knew when not to complain, but he clearly loved her in his own way, and he missed her when she was not around. For her part, as her birth was illegitimate, this suggests a vulnerability that may have inclined her towards involvement with a strong — and legitimate — protector/father figure. They suffered the death of a daughter together and she supported him through most of his difficult times, though not quite all. However, the marriage has inspired some apocryphal tales because it was seen by others to have been very one-sided. One anecdote has it that on the night of their wedding, as they prepared for bed he told her, 'Now Clementine, I want you to know that I love you. I have always loved you, I married you because I love you, and I always will love you. Now please let me hear no more about it'. Of this marriage, Violet Asquith said:

> His wife could never be more to him than an ornamental sideboard … and she is unexacting enough not to mind not being more … he did not wish for — though he needs it badly — a critical reformatory wife who would stop up the lacunas in his taste and hold him back from blunders. (Jenkins, 2001, p. 138)

Churchill had several successes but also many failures in his early years in politics. His greatest success came after he abandoned the Conservative Party and joined Lloyd George's Liberal government to help set up the British welfare system, to aid widows and the needy, to fund unemployment insurance, to develop the British system of labour exchanges, and to pass the Trades Board Bill, all of which drastically improved conditions for the working poor. He also supported the vote for women. In undertaking these reforms, he told parliament, he was moved by the debt he felt towards his childhood nanny, Mrs Everest, whom he fondly visited in her little shack near the end of her life. But his sympathies were patrician; he did not pretend to understand the underprivileged, he merely sympathised with them from on high as if he were of a different order or race. He championed the underdog, particularly against the middle dog, but only on the understanding that he was top dog.

His greatest failures came during World War I. He had previously gained a reputation for incompetence over a brush with anarchists in 1911, and when that same year he had used 50,000 troops to crush a railway strike, he became deeply unpopular with the public. During the war he was blamed for the losses of the Battle of Coronel where 1,600 sailors died, and the failure to save Antwerp. He was blamed for the sinking of the *Lusitania* with the loss of 1,198 lives, and certainly he should have prevented it from sailing. Then he embarked upon the Gallipoli invasion, a failure for which he was demoted within the Cabinet. His arrogance and refusal to apologise further outraged the public against him. When a report into his failures only partly exonerated him he attacked it vociferously, churlishly casting aspersions at its author who, because he had unexpectedly died on the eve of the release of the report, was unable to defend himself, although the report was very fair to Churchill. This made him even more unpopular.

He may have become suicidal at this point; he wrote a letter to Clementine outlining the terms of his will, and he added, 'Don't grieve for me too much. Death is only an incident'. For therapy he took up painting, which he found he excelled at and which he continued to enjoy for the rest of his life. But on one occasion, having spilt red paint on himself, he looked around in agony and exclaimed, 'There is more blood than paint on these hands. All those thousands of men killed …'

He was shunted sideways to become the Minister of Munitions, where he championed the innovation of the caterpillar tank, without which World War I would not have been won. Then he angrily resigned his portfolio and returned to his old unit in the trenches where he took command with the pronouncement: 'Gentlemen, I am now your Commanding Officer. Those who support me I will look after. Those who go against me I will break'. Yet he came to be loved by these same men.

Upon re-entering parliament he helped launch the War of Intervention into Russia, in which a misguided alliance of several European nations and the United States invaded Russia, and actually came to occupy a large amount of the Russian landmass at one point. This war pitted White Russians against Red in an adventure that Churchill hoped would see the war-weary English rally to the flag once more, and rise up to crush the Bolshevists, thereby satisfying his need for applause. When they did not he found himself even more unpopular. Yet he continued to promote Jennie's progressive political agenda, including social welfare and prison reform.

He rejoined the Tories in 1925 and was given responsibility for the Air Force. This inspired him to attempt to get his pilot's licence, which resulted in a crash and another close call, neither of which dented his enthusiasm. Eventually, Clementine dissuaded him from continuing this quest. However the parliamentary work he was assigned to seems to have led him to feel that his best years were behind him by age 40 (his father died at age 46). He carried on his journalistic work at the same time and steadily built up a fortune (his parents had squandered their inheritances) until the Wall Street Crash nearly wiped him out. He spent most of the 1930s in the (relative) political wilderness, developed depression — his famous 'Black Dog' — and drank even more than usual. On one occasion he unleashed an invective against Gandhi, who was then campaigning for Indian independence, that bordered on the racist and exposed his aristocratic elitism, for he was really a 19th century man of the Empire rather than a progressive modernist.

However, like many charismatic personalities, he seemed able to recreate himself in mid-life. He perceived the threat of Hitler as early as 1933, long before his political colleagues and, in the face of this threat, he became a changed man, taking a stand from which he refused to back down. He moderated the extremes of his personality and focused on the single mission of saving Britain and democracy from Fascism. He was extraordinarily prescient as to what might be the most likely course of events, at one point composing a memorandum that postulated full-scale war with Germany, and even giving a timetable for this that was remarkably detailed and accurate. He was no warmonger but he campaigned constantly for the re-arming of Britain. He also seems to have had an uncanny insight into international affairs; when he was a young cavalryman serving in Cuba he had correctly foreseen the Spanish–American War; unfortunately his seniors in Parliament and the Defence Ministry ignored his memorandum. In 1937, he also wrote an essay that accurately predicted the creation and problematic future of Israel.[2]

In time, Churchill went on to become Britain's greatest ever wartime leader, growing far beyond his father's shadow. Again, like most charismatic personalities, he was at his best when rising to meet a challenge. However, though brave and mostly competent he was not a great military man; his foray into Norway during World War II repeated his failure at Gallipoli, and he had no deep military insight. Rather, his genius resided in his strategic vision — recognising the danger of the Nazis — and his inspirational leadership. He made many mistakes in lesser matters, but he got the one great matter right. Any other British politician of the time would have made peace with Hitler, thus colluding with Nazism and enabling Germany to conquer Russia, but Churchill stood firm. He exploited his brilliant rhetorical ability to its greatest effect, devising his greatest phrases about 'blood, sweat and tears', telling of 'their finest hour', and of 'so much owed by so many to so few', and especially vowing that 'we shall fight them on the beaches', leading to one commentator saying that he marshalled the English language and sent it in to war (Hume, 1994).

Although his judgment was flawed — he was conned by Stalin just as Lloyd George had been fooled by Hitler (and Stalin, that most suspicious of men, also utterly misread Hitler) — his courage was unquestionable, especially during the years 1939–1941 when Britain stood alone against Nazi Europe, withstanding the Blitz and fighting in Europe, Africa and Asia. In May 1940, after only a fortnight as Prime Minister, and against the urgings off his inner Cabinet, he refused to negotiate with Hitler when all seemed hopeless (Donaldson, 2003). He did not shrink from the terrible tasks imposed by war, such as the sinking of the French fleet at Oran, and the carpet-bombing of German cities. (This last action was undoubtedly the most controversial of his life, although David Irving's mischievous estimate of 250,000 dead at Dresden has now been corrected; probably about 25,000 died, and Churchill did cease such operations after a few weeks.) He also took to visiting the sites of bombed-out English towns and suburbs and freely expressing his feelings about the survivors and their plight, leading one woman victim to declare, after noticing him moving among the survivors with his eyes filled with tears, 'Look, he really cares'. (A film of this episode had an inspiring effect on Nelson Mandela many years later.)

Despite his inspiring wartime leadership, Churchill received a rude shock in the first postwar national election held in June 1945, when he led the Conservative Party to one of its worst-ever defeats. The reason was that although he was personally massively popular, he was perceived by the electorate to be a wartime leader and the populace was eager to put the war behind them. The voters had no sympathy for his continuing warlike rhetoric in peacetime, which included likening his Labour Party opponents to the Gestapo, and they feared his aspirations to contain communism and to re-establish the Empire. They wanted a man of peace rather than a warrior, no matter how inspiring. In peacetime they were intolerant his less-savoury side, for Churchill could also be mean and small; in 1942 he stopped Noel Coward's knighthood (it had already been approved by the king) because of jealousy and a touch of homophobia.

Undismayed by defeat he returned to his writing, turning out *The Second World War (1948–1954)*, and *A History of the English-Speaking Peoples (1956–1958)*; books that, like most of his output, are effective but not great. He regained the government benches in 1951 at age 76, but by now he was unreliable and indifferent to his workload; he had become increasingly so since the last months of the war. He suffered a stroke in 1953 from which he never fully recovered. He clung to power but he was eventually ousted as Prime Minister in April 1955, although he remained a member of Parliament for several more years. He accepted retirement reluctantly, suffered another stroke in January 1965, and died soon after.

Any evaluation of Churchill has to recognise first that he lived in different, and much bloodier, times and thus some of his more questionable actions were seen, both by him and others, as merely what had to be done in difficult circumstances. Second, his cruel early life experiences imposed limitations on him that he could never hope to escape. Actually, Churchill's harsh upbringing was merely a variant on the time-honoured way that many of the ruling class raise their male children (Kershaw, 2007), but whereas many emerge morally and socially blunted from such experiences, a few transcend them and go on to lead remarkable lives, albeit scarred by their sufferings. Churchill did both, and thus any success he had was relative. Hence the ambitious busyness that he maintained throughout his life was highly neurotic, and was also a flight from depression. When someone stays as busy as he did it is likely to be because they are frantically doing something that they are successful at (or hope to be) in order to fend off depression arising from the crippling effects of childhood abuse or neglect. By most of the norms of mental health Churchill was severely psychologically damaged, and any judgment on him must take account of this also.

The underlying theme through all of Churchill's life was his yearning for love from his father. There are many poignant accounts of him pining for his father's approval. His friend and biographer, Violet Bonham Carter, has recalled how, after a conversation she had with him about politics, he asked with wistful envy, 'Your father told you that? He talks about things quite freely? I wish I could have had such talks with mine ... But I should have had them if he had lived. It must have come' (Carter, 1965, p. 23). We can read through such accounts that Churchill was at great pains to deny his parents' indifference to him, bluntly, their lack of love. We can also deduce that he probably felt a towering rage at such rejection. Curiously, he may have fallen into the same trap he was trying to transcend with his own son, whom he named Randolph after his father, but from whom he became estranged.

In 1947, after his wartime success and at the age of 73, he had an experience that he called 'The Dream' (it is unsure whether it was a dream, a daydream or a fantasy) that he later recorded for posterity. As he told it, one day when he was painting (this had become his favourite hobby) he felt an odd sensation. Turning around, he saw a vision of his father standing nearby and looking just as he had over 50 years previously. They conversed and Winston explained to his incredulous father the many changes that had taken place in Britain and the world over the previous half century. His father seemed unaware of how Winston had risen in the world, assuming his son was a retired soldier or a professional artist. At the end of this discussion his father told him,

> As I listened to you unfolding these fearful facts you seemed to know a great deal about them. I never expected that you would develop so far or so fully. Of course you are too old now to think about such things, but when I hear you talk I really wonder [why] you didn't go into politics. You might have done a lot of help. You might even have made a name for yourself.

Before Winston could offer another word of explanation there was 'a tiny flash' and his father disappeared.

It seems that Churchill's entire life script was an effort to get love from his uncaring parents. His natural genius was painfully conflicted by his parents' indifference to him, yet it was also stimulated by necessity to extraordinarily high levels of creativity, probably much higher than he would have attained had he not been so desperately driven. In order to endure, he fashioned a myth that cast his parents as loving but indisposed. His parents were great but he could be great too, and then they would have to give him their love ('It must have come'). As a small child living in his world of toy soldiers he rehearsed a future destiny in which he would achieve great things and thus earn his father's love and respect. His parents' indifference severely damaged him; he was plagued by both depression and alcoholism throughout his adult life. But near the end he seems to have felt that he had acquitted himself, to have achieved some awareness and some resolution of his emotional vicissitudes, to have fought his demons to a standstill.

Unusual among charismatic individuals, Churchill could be reflective and introspective. When World War I was declared he was elated that his opportunity for greatness had at last arrived, but in the middle of his joy he suddenly turned to his wife and inquired about this elation, asking 'Is it not strange to be so constituted?' Hence it is likely that, in some corner of his mind, Churchill understood the terrible tragedy of his life, the unrequited yearning and his pursuit of his parents' love, and in his 'dream' he achieved what peace he could. It did not silence his yearning for parental approval, for even in his final years he still considered himself a failure — he had not restored Britain to its former glory, indeed, he had witnessed the dismantling of its empire. But he seems to have felt that had his father known of his achievements he would have loved him; he felt that he had proven himself worthy by any criterion. His last recorded statement (made in January 1965) was, 'It's been a grand journey, well worth making once'. He died at the age of 91.

Violet Bonham Carter has written a moving account of him. She experienced a typical charismatic affection for him, describing her feelings for him as, 'This is what people mean when they talk of seeing stars' (Carter, 1965, p. 16). She has described his quite extraordinary qualities, his unabashed confidence, his unquashable resilience, his dash and flair, and his unusual mental processes, processes that indicate that charismatic personalities experience the world differently from others. As well as the quote that heads this chapter she has written:

> To Winston Churchill everything under the sun was new — seen and appraised as on the first day of creation. His approach to life was full of ardour and surprise. Even the eternal verities seemed to him to be an exciting personal discovery (he often seemed annoyed that some of them had occurred to other people long ago). And because they were so new to him he made them shine for me with new meaning. However familiar his conclusion it had not been reached by any beaten track. His mind found its own way everywhere ... His world was built and fashioned on heroic lines ... Although he had the ageless quality of greatness I felt that he was curiously young. In fact, in some pedestrian ways he made me feel that I was older. I felt that, though armed to the teeth for life's encounter he was strangely vulnerable, that he would need protection from ... a humdrum world which would not easily apprehend or understand his genius ... He was as impervious to atmosphere as a diver in his bell ... Winston Churchill possessed a power of concentration that amounted almost to an obsession. It gave his purpose a momentum which

often proved irresistible. But the rock on which that purpose sometimes foundered was the human element he had failed to take into account. Though he was the most human of all human beings he was himself far too extraordinary to know how ordinary people worked ... Even in those early days he felt he was walking with destiny, that he had been preserved from many perils to fulfil its purpose ... (Carter, 1965, pp. 18–23)

Carter added that Churchill had such a rare and subtle turn of mind as to comfortably combine grandiosity with self-deprecation. In what is perhaps the best summation of him she recalls him telling her, 'We are all worms, but I believe I am a glow-worm' (Carter, 1965, p. 16).

Endnotes

1   The primary source for this chapter is Jenkins (2001).
2   Cited by Robert Fisk in the program *Big Ideas*, introduced by Phillip Adams on Radio National, Australian Broadcasting Corporation, March 26, 2006.

CHAPTER THREE

• • • •

The Psychoanalytic Perspective

A man's work is nothing but a long journey to recover the two or three great and simple images which first gained access to his heart.

Albert Camus

Winston Churchill's life demonstrates many of the recurring themes of charismatic personalities, including a problematic childhood, extraordinary talents such as his phenomenal memory and inspirational rhetoric, unusual mental processes and a need for fame and applause. He also sought out problems and conflicts so as to present himself to others as a problem-solver. He had the foresight to marry an undemanding partner, and the capacity to recreate himself in mid-life. Underlying these, however, was a pervasive narcissism that was both a strength and weakness. This narcissism was his adaptation to his difficult childhood, yet it stimulated him to develop to extraordinary levels those native talents he possessed. This process of adaptation and stimulation transformed him from a tragic misfit into a charismatic personality. To begin to understand all of this, Ovid's myth of Narcissus is a natural starting point for a theory of charisma.[1]

Narcissus, an extraordinarily beautiful young man, was the product of the rape of his mother Liriope by the river god Cephisus; hence he was part human and part god, conceived in trauma and born into ambivalence. So beautiful was he that envious gossips went to Tiresias, the only person to have lived as both male and female and who had been given the gift of prophecy by Jove, to ask whether anyone so beautiful could live for long. Tiresias answered enigmatically by saying that yes, Narcissus could live long, 'unless he learns to know himself'. This is generally interpreted to mean that provided the narcissist remains within their world of self-love, then their narcissism can live long. However, should they learn of their condition then their self-absorption will be short-lived. For although the narcissist thinks only of himself, ironically he does not really know himself because he is unable to see himself objectively, that is, as others see him, or as he really is. To do so is to bring his world crashing down.

At 16, Narcissus had many admirers but he abstained from human relations. Then one day the wood-nymph Echo, herself a cursed being, fell utterly in love with him. Echo's plight had been to be rendered unable to speak anything but the last couple of words said by another; she had no ability to speak for herself. This is generally interpreted to mean that she represents a mirror image of the psychology of Narcissus; whereas he is able to think only of himself, she was unable to express any thought of her own. Thus, whereas he was self-absorbed, she virtually had no self. This was illustrated when Narcissus became lost in the woods and called out to his friends, 'Come to me'. Hearing this, Echo revealed herself and repeated, 'to me … to me …'. But Narcissus ran from her saying, 'I would rather be dead than let you touch me'. At this Echo withdrew, humiliated, to hide in the forest. Her bones and body shrivelled up and turned to stone, and her voice roamed off by itself to live in caves and canyons.

Resentful at rejection and at the fate of Echo, one of Narcissus' would-be lovers lifted his hands to the heavens and prayed, 'Let Narcissus love and suffer as he has made us suffer. Let him, like us, love and know it is hopeless. Let him, like Echo, perish of anguish'. Nemesis, the Corrector, heard this plea and granted it.

One day, thirsty from hunting, Narcissus bent to drink from a pool of perfect water', but as he did so he saw his reflected image and fell deeply in love with it. Of course, all his efforts to embrace and kiss his image left him frustrated and lovesick, and eventually he recognised 'You are me, now I see that … but it is too late. I am in love with myself …'. He realised that there could be no solution to his predicament, and that he must die: 'I am a cut flower … let death come quickly'. The tragedy of his situation brought him to his first realisations of empathy and compassion towards others, though still expressed in terms of concern for himself: 'The one I loved should be let live. He should live on after me, blameless'. However, he knew that this was impossible, and when he died both his narcissistic ego and his observing self died also. Yet even as he crossed the river Styx into Hades he could not resist getting a final glimpse of his reflection in the water. At the moment of death however, he was transformed into a beautiful flower, the Narcissus, with its seductive fragrance and its delicate evanescent trumpet.[2]

Havelock Ellis, the 19th century sexologist, first linked the myth of Narcissus with psychopathology. Ellis viewed homosexuality as a pathology of self-love, and a reviewer of Ellis' work, Wilhelm Nacke, used the term 'narcissism' to refer to morbid self-love. In his 1914 paper 'On Narcissism', Sigmund Freud adopted the concept and advanced a theory about it.[3] More recently, contemporary psychoanalysts have renewed the study of narcissism (neglected for decades). There has been a certain romantic wildness to some of their theorising; one worthy holding that narcissism is responsible for cancer, endometriosis, low-back pain and schizophrenia among other ills (Symington, 2002, pp. 1–2). However, more thoughtful theorists, specifically Heinz Kohut (1971, 1977) and Otto Kernberg (1975), have also presented detailed and thorough accounts of narcissism from an analytic perspective. Kohut's theory is a departure from mainstream analytic theory, but because he applied it (partially) to a charismatic personality (Kohut, 1976) it will be used herein, with brief concluding comments and a table illustrating Kernberg's approach is also included.

Kohut observed numerous similarities between his narcissistic clients and charismatic leaders. Unlike most therapy clients who present in a demoralised manner, his narcissistic clients presented with grandiose self-confidence and an extraordinary lack of self-doubt.[4] They were often quite clear-headed and perceptive, persuasive and accusatory; Kohut has recounted how one such patient accurately diagnosed his (Kohut's) shortcomings while in therapy. However, with time this façade of strength and competence gave way to a vain boasting and a naïve sense of omnipotence and invincibility. Grandiose fantasies and exhibitionistic themes appeared in their conversations, and so 'brittle' did their confidence and self-certainty become, so desperate were they to impress, that they were sometimes unable to admit to any gap in their knowledge or to ask for routine information or advice. Often there was a feeling of a lack of any real contact or dialogue with these clients; they might appear to listen politely then immediately return to their own preoccupations, as if what had just been said was of no consequence. They seemed extraordinarily self-contained, revealing little or no discernible impact upon them by the therapist. They had been reluctant to seek therapy but had been obliged to do so because they had been compromised by various fraudulent or sexually perverse behaviours. As therapy progressed they became increasingly hypochondriacal and self-pitying, and the nearer Kohut got to the core of their disturbance the more catastrophic were their reactions. They demonstrated little or no conscience or guilt, and they showed in their relationships that they basically saw others merely as extensions to their own egos.

In sum, narcissistic patients appeared to be functional, happy and healthy until one looked a little deeper. Then a profound emptiness was revealed, an emptiness that had been split off from the personality and suppressed from awareness by a powerful 'all-or-nothing' streak that was utterly committed to self-control and an appearance of strength at any cost. Their extreme autonomy and their superficial charm made them quite appealing to others, who warmed to them, as if recognising some part of themselves in these figures. This 'mirroring' process, in which a strong figure uses others as parts of their self while these others live through him or her, alerted Kohut to a narcissistic explanation of charisma.

According to psychoanalytic theory, narcissism begins with the sense the baby has of being the centre of the universe. In the womb the baby develops a feeling of oneness; it has no sense of any 'other' or even of any 'otherness'. It just is, secure and total. As far as it knows, it is all that exists, for without the experience of another, the possibility of another's existence never arises. Significantly, even after birth the infant may still experience the mother's body — where pleasure and sustenance come from — as part of itself.

In this early period of 'oneness' (Mahler, Pine, & Bergman, 1967), a dialogue of mutual cueing and empathy, of quiet gesture and moulding, develops for both mother and infant. Through the mother's holding and feeding, a choreography develops in which their boundaries seem to melt away (or, more accurately, the very discovery of boundaries is delayed for the baby). It is as if both mother and child merge, the being of one melting into the being of the other (Kaplan, 1979, p. 100). With such close empathy the child perceives the mother as merely part of what is; that is, as a part of itself.

The infant now feels exalted in its mother's eyes, omnipotent in its childish world and grandiose in its uninhibited egoism. The baby is a conqueror who seemingly creates magic without understanding how or why. The infant is the recipient of devotion and unconditional love from the mother. It feels special and even magical (it only has to cry and food appears). The rising nipple finds the hungry mouth, and a warm yielding softness that feels and smells just like the child moulds itself around him. From this comes the illusion that his feelings and gestures have created the nipple, the mother's body and the rest of the world (Kaplan, 1979, p. 92). The child feels like 'an angel baby held in the sumptuous lap of a saintly Madonna' (Kaplan, 1979, p. 116), his love (self-love that is, for without some concept of an 'other' only self-love can exist) arising from a sense of shared perfection with the mother, who is experienced as a part of himself.

In normal development this 'primary narcissism' as Sigmund Freud called it (Freud, 1914), evolves into more mature forms of narcissism such as self-esteem, creativity, idealism, conscience and other traits in accordance with the Reality Principle (another term from Freud). This involves the developing child gradually learning that it is not the centre of the universe, only of its mother's attention, and that sometimes even she is absent or uncaring, and that the external world — the 'real' world — contains many painful surprises; that most people care little or nothing for the child. Through the reality principle the child learns skills such as reality perception, frustration tolerance and impulse control, and also to develop secure ego boundaries.

This occurs through 'optimally failing parents' (Kohut, 1977, p. 237), and optimal frustration of the infant in a secure family environment, where the mother can coach her child towards separateness and autonomy. The gleam in the mother's eye mirrors the infant's exhibitionistic display, and her participation in the child's egotistical enjoyment confirms the child's self-esteem, despite whatever painful encounters with the world occur. By gradually increasing the selectivity of her responses the mother channels the baby's behaviours in realistic directions, and the sense of oneness slowly breaks down (Kohut, 1977). This, in turn, leads to the consolidation of the child's self. The store of self-confidence and self-esteem that sustains one through life derives from those early difficult but ultimately successful (if given loving support) struggles. Any pathological elements (of which there are bound to be a few) gradually become integrated into the self, to perform realistic functions in the service of the ego, as a functional and prosocial inner balance is achieved.

However, for some this development goes awry, and Kohut gives us two scenarios for how this might happen. The first involves actual damage to the psyche, usually because of underinvolved or inconsistent parenting, and an effort by the child to repair the damage with a grandiose fantasy. The second results from a specific failure to mature emotionally past a particular stage of development, with resultant regressive fixation and an inability to have certain kinds of feelings and thoughts, usually because of overinvolved parenting.

In the first scenario, the infant is inconsistently and unpredictably related to by its caregivers in alternatively loving, rejecting and indifferent ways, or there may be significant abuse, neglect or conflict in the family-of-origin. These disruptions are usually not quite so severe as to traumatise or impair the child (although they might), merely to incline him towards narcissistic thought processes and behaviours as a compensation for familial deficits, and as a defence against feelings of emptiness, unreality and unwantedness. Because the child has no way of knowing how it will be treated next, and because such unpredictability and inconsistency are intolerable, a child who is exceptionally talented and adaptable may make a heroic effort to rationalise such disruptions in a comforting way. It may develop a protective myth that explains the mother's absences or lapses of rapport, or even conflict or neglect, and that sustains the child through them. This myth maintains the child's sense of special value and is reinforced by those times when it is treated in a loving manner. Such loving moments — for example, reunions with a neglectful parent — become proof that the myth is true. Living within this myth, the child constructs ways of excusing, diminishing or ignoring those times when it is treated indifferently or rejected. It comes to 'know' it is special and dismisses or denies or reinterprets evidence to the contrary. This seems to be what happened in Churchill's case.

This scenario is likely to become suffused with rage. The unpredictability of the absences or failures of rapport, the painful feelings of loss that they produce, and the work that the child has to do to maintain itself during them, cannot but provoke intense anger. But it can be very scary to express rage against a beloved person, or even to feel it, especially if the relationship is basically insecure. It can be done, but to do it well usually requires optimum conditions (such as a longstanding relationship between mature equals as in a good marriage), and it invariably results in much guilt, perhaps more than a young child can stand. If there is any likelihood that the feeling or expression of such rage or guilt may cause another failure of rapport, or may even cause the beloved to abandon the child (and such a thought has to be considered likely to occur to the child, and to terrify it), then the rage will be suppressed.

In the second of Kohut's scenarios there is little or no significant abuse, neglect or conflict in the family-of-origin. Rather, an extremely devoted and idealising mother whose 'baby worship' has already created a child with very high self-esteem (Freud's famous phrase, 'his majesty the baby' expresses the sense of this), maintains her intense involvement long past the time when primary narcissism would usually break down. What likely happens is that an anxious, insecure, unhappy or driven mother pours all her hopes and dreams, disappointments and frustrations into her child, praising him, stimulating him and ensuring that he always feels especially loved. Certain primitive qualities of infancy — oneness, grandiosity and exhibitionism for example — are supported by the mother for much longer than they ought to be for optimum development.

The effect of this is to educate the child to naturally believe that the entire world will relate to him as his mother does. He comes to expect to be recognised as special, indeed, to need and demand it and to recruit others into relationships with him that will recapitulate the old maternal dynamics. Lawrence Olivier described this in his *Confessions of an Actor*, writing that, during his childhood, as long as his mother was home he never 'played to an empty house'. The early maternal relationship becomes the matrix for all subsequent relationships; it becomes the only kind of relationship in which he can feel fulfilled or secure. Later relationships are mapped onto it and where they are found wanting they are either changed or abandoned. Eventually, he becomes incapable of less elevated relationships and life becomes a quest to convert realistic relationships into more worshipful ones.

Such a scenario also becomes suffused with rage. The child cannot help but feel rage towards its mother for being so overinvolved, although it does not understand what is happening and its contacts with its mother are invariably 'loving'. Of course, it is not genuine love to be so overprotective of a child when it ought to be learning to protect itself, but this may be the only kind of love of which the mother is capable. Further, when the child does enter into the world of other children and parents it finds that they do not conform to its expectations as its mother did, and this too elicits rage. But it is an odd kind of rage that develops; a subterranean rage, for the child has never had the opportunity to feel rage at its mother, so close and comprehensive has her attunement and protection been. Hence, such rage is likely to be kept within superficially friendly relations. In sum, it is a repressed rage that is never admitted consciously, and which gets covered over by apparent warmth and normality, yet which becomes structured into the developing personality.

Several cautions need to be made about all this. First, it is not usually the effects of trauma and abuse that are most damaging to a developing child, nor is it parental attitudes of exploitation, seduction and deception, painful as these all may be. The greatest damage wrought upon developing children comes more from the background emotional climate, especially a sense of being unloved or unwanted, or of being loved only conditionally, or the unpredictable withdrawal of love (Millon, Millon, Meagher, Grossman, & Ramnath, 2004, p. 93). In considering Kohut's scenarios it is important not to seek some single event to explain narcissistic development, but to consider the underlying relationship qualities and the messages the child receives about its value and place in the world, for this is where the child is most vulnerable.

As well, it is important to recognise that many other developmental possibilities may arise from these scenarios of parental over- and underinvolvement, ranging from collapse into mental illness, drug dependence, compulsive promiscuity or other antisocial acting out, to sublimation into art, ambition or the priesthood. All of these possibilities and more occur frequently, but they would seem to be less likely to result in the development of a charismatic personality. Probably less than one in one hundred potential candidates for charismatic development eventually attain it. Quite why children from very similar backgrounds develop differently is seldom clear, and may involve subtle differences that are beyond the current ability of psychology to measure, though it happens all the time.

Both of Kohut's scenarios are also quite 'normal', albeit they are distortions of a normal process. In the first, the normal 'optimal failure' of the underinvolved or inconsistently involved parents gets magnified to intolerable levels, forcing another quite normal process, the myth-making that we all do, into narcissistic directions. In the second, there is a selectively atrophied maturation resulting from the buffer from the reality principle that is provided by an overinvolved parent; the child continues to grow and to develop, but certain emotions and relational possibilities are constrained. What both scenarios have in common, as well as the strong feelings of stifled rage that they produce, is that they may stimulate such a child to make a leap of development that would be impossible for an ordinary child, but may just be possible for a very resourceful child in a conducive environment. Such a leap would bypass the reality principle and deny and avoid the problems encountered, while appearing to usher in normal functioning. This occurs when the child takes onto himself his parent's 'filter' mechanisms.

Normally, the mother filters the environment so that the child is not exposed to unpleasant events beyond its ability to cope. When painful things happen she interprets and evaluates them for the child in a positive manner. When absences or failures of rapport occur, or when the child is cast into the external world after a long period of overprotection, it may seem quite natural and automatic for it to mimic the mother's interpretive and evaluative strategies, to incorporate her techniques as part of its own self. Thus the child brings adult skills to bear on childhood adaptive problems (intriguingly, charismatic personalities have often been described as combining an ageless or 'eternal' wisdom with childlike innocence). Hence, rather than falling from grace with her, the child identifies with her even closer (emotionally), and retains or even increases the primitive sense of oneness that sustains its grandiosity.

Now, instead of surrendering his narcissism the child draws on all his resources. He learns to charm, manipulate, bully and calculate his way through situations that defeat others. Alone he denies his aloneness and defies the world, yet without understanding the significance of his actions. As one who refuses to grow up, or perhaps is unable to, the child manages to avoid the reality principle — compromise with an indifferent and dangerous world — and his egocentric view of life remains substantially intact. As part of this, Kohut says, he becomes 'superempathic' with his self and his own needs.[5] The result is a remarkable autonomy in which the narcissistic child asserts his own perfection, yet uses others to regulate his self-esteem, demanding control over them without regard for their rights as independent people. This leads to a severe reduction in the educational power of the environment (Kohut, 1976, pp. 424–425). The narcissistic child lives in a psychological world of his own creation, beyond or outside of 'normal' reality, and virtually unreachable at depth.

Such a person grows up behaving normally, for he has learned the appropriate behaviours to get rewards and avoid punishment. He may even come to project a commanding substantiveness beyond his years because of his incorporation of aspects of the parental mentality. But deep down he still views the world as an extension of his self in the way he originally saw the mother, and his early relationship with her remains the model for all subsequent relationships. Perhaps we all do this to some extent, and have varying degrees of insight into ourselves, but for the narcissist these insights remain purely intellectual. Deep down he 'knows' the world revolves around him, and his adult life is an attempt to perpetuate his childish egocentrism.

The forms and transformations of narcissism vary greatly. Sometimes there are pathological elements, but always the result is a person who sees the world in a radically different way from others. He is likely to be enormously confident and fearless, for how can one be afraid of anything in a world that is felt to be an extension of oneself? It would be like being afraid of one's own leg. But while he may seem superficially to be a product of the society he grew up in, really he is his own universe, and only his body, needs, thoughts and feelings are experienced as truly real. Others are perceived intellectually but without emotional weight and colour; without substance or relationship and without conscience. Perhaps they are experienced in a manner analogous to how we experience our internal organs; we know they exist and are real, sometimes we even worry about them and they give us pain, but we never actually encounter them.

Such a person grows to be emotionally detached from the real world, although his world is real enough to him. By always being a little bit inside yet a little bit outside the world, he is well placed to diagnose its problems and devise solutions. He is elevated above commonplace humanity and this elevation gives him an overview and a cool analytic distance. His diagnoses and insights will seem to be profound truths to those who share his values or social location. The talents that he has developed to survive with his narcissistic worldview intact may now give him an uncanny resonance with his times, and the needs and hopes of those who will become his followers. In addition, other key traits associated with childhood (to be discussed below) may be developed to an extreme because of their survival value. Perhaps this really does give the charismatic personality a 'truth' that others lack.

Eventually, the successful adult narcissist stands ready for the call to leadership. He fits in well with those who seek a new life or are in crisis. In return for their love and devotion he leads them in the fulfilment of some great project they desire, and in so doing he recreates for himself the ego-reflecting universe he knew as a child. As Erik Erikson put it, he 'solves for others problems that he has been unable to solve for himself'. The followers appreciate his vision and wisdom because he keeps his head in a crisis; he is above the fray (which is why narcissistic patients are so hard to treat — they cannot be reached [Kohut, 1976, pp. 414–415]).

Kohut gives us two images of the charismatic personality. In the first and least flattering he discusses Wilhelm Fliess's relationship with Sigmund Freud (Kohut, 1976). Fliess was an eccentric and incompetent surgeon whose writings and numerological speculations, especially his 1897 book *The Relationship Between the Nose and the Female Sexual Organs from the Biological Aspect,* have been variously described as 'Teutonic crackpottery' and 'disgusting gobbledygook'. He came up with a theory of infantile sexuality and a surgical 'cure' for masturbation (an operation on the nose) from which Freud drew extensively in his own theorising. But Freud seems to have experienced a charismatic affection for Fliess, assuming an attitude of reverence and submission towards him, even asking him to perform an operation on the nose of his patient Emma Eckstein to cure her of her 'nasal reflex neurosis' (a nonexistent complaint) that he believed was caused by masturbation. Fliess bungled the operation by leaving a wad of gauze in Eckstein's nose. This might have cost her her life, had not another surgeon intervened. Yet Freud made excuses for his friend by arguing that Emma's haemorrhaging was actually hysterical and had been motivated by an unconscious desire to entice him to her bedside.

Freud's early biographer, Ernest Jones, has explained Freud's attachment to Fliess by noting that Freud used Fliess as a catalyst to enable him to liberate his speculative side. During Freud's most creative period, his self-analysis that preceded his greatest work, *The Interpretation of Dreams* (1900), he became weak and needy in the throes of a supreme creative effort. Many thinkers and artists need emotional support during such periods of intense creativity, especially when their exploratory forays lead them far beyond familiar frontiers into lonely areas not previously mapped by others. Such isolation can be terrifying because it repeats an early childhood fear of being alone and abandoned. As Freud wrote during this time, 'I live in such isolation that one might suppose I had discovered the greatest truths' (Kohut, 1976, p. 402). At such times Kohut remarks that even a genius like Freud may attach himself to someone whom he sees as wise and powerful, and with whom he temporarily bonds in order to draw creative support. The bombastic Fliess, and others like him with their unshakable self-confidence, certainty and wild ideas, lend themselves to this role (Kohut, 1971).

Freud corresponded with Fliess (they lived in different cities) weekly for several years while undertaking his self-analysis. Freud deferred to Fliess's theories and used him as a sounding board for his own speculations. With the aid of such communication, encouragement and mutual admiration, Freud was able to sift through his own ideas until he eventually arrived at the theory of the mind that he elaborated in The Interpretation of Dreams. However, although Freud never renounced Fliess — indeed, he seems to have held a high regard for Fliess's theorising for the rest of his life — he did abandon him soon after he had finished his self-analysis, and he later edited him out of the history of psychoanalysis. The relationship had a purely functional purpose, despite the fact that at times Freud looked on Fliess as a messiah, and described their correspondence as 'the most intimate you can imagine' (Webster, 1996, p. 218).

The Fliess type is similar to charismatic leaders like Hitler and Napoleon who have enormous, yet brittle self-esteem (Strozier, 1980). Lacking self-doubt they set themselves up as leaders. Their unshakable self-confidence makes possible great leadership but also risks total failure, for such people lack flexibility and have an all-or-nothing quality with only two options: success through strength, or destruction through defeat, suicide or psychosis (Kohut, 1976; Strozier, 1980). Such leaders may be quite paranoid, but what fits them for their role is the fact that their self-esteem depends on their incessant use of certain mental functions. They continually judge others and point out their moral flaws, then — without shame or hesitation — they set themselves up as leaders and demand obedience to 'improve' those whose flaws they have diagnosed. They typically have no sense of guilt and are highly sensitive to slights — even somewhat paranoid — while also being adept at turning such slights to their advantage by evoking guilt in others. They have an unusually strong ability to maintain their self-esteem, for without such strength they would crumble. They have stunted empathy for others because they have emotionally withdrawn into themselves, with the result that they are 'superempathic' with their own emotions; they can listen to their feelings well, although they have little genuine understanding of themselves. They are also fundamentally hostile, enraged at a world that ignores their narcissistic claims; Kernberg called this 'oral rage' because its origins lie in the very earliest stages of development (Kernberg, 1975, p. 228). Yet they depend on their followers; Freud took up with Fliess when he needed him but dropped him soon afterwards, and some of Kohut's patients behaved similarly (Kohut, 1976). Kohut insists that although such relationships are opportunistic they are not pathological. Anyone who is in need of support, for whatever reason, may be drawn to charisma (Kohut, 1976).

Charismatic personalities come in all shades and degrees. A few are almost psychotic; dogmatic, blind fanatics whose only talent is an unusual cunning. But others are quite different, as Kohut illustrates in a discussion of Winston Churchill, his second example. Intriguingly, Kohut has speculated that Churchill may also have retained the infantile delusion that he could fly; the language and style of *My Early Life*, his escape from the Boers and his leadership of Britain during World War II during its 'escape' from the Nazi threat, all suggest this. As Kohut notes: 'Churchill repeated again and again in an ever-enlarging arena, the feat of extricating himself from a situation from which there seemed to be no escape by ordinary means' (Kohut, 1966, p. 256). Churchill's wild leap from the footbridge during adolescence is another indicator, about which Kohut adds:

> On this occasion the driving unconscious grandiose fantasy was not yet fully integrated (although) the struggle of the reasoning ego to perform the behest of the narcissistic self was already joined. Luckily for him and for the forces of civilization, when he reached the peak of his responsibilities the inner balance had shifted. (Kohut, 1976, p. 257)

Kohut explains the significance of flying fantasies as: 'Dreams of flying are an extension of the aspirations of man's grandiose self, the carrier and instigator of his ambitions' (Kohut, 1977, p. 113).

Between the extremes of Fliess and Churchill there are many possible types of charismatic personalities. Their common features are an extreme narcissism in which they identify others, and perhaps even the entire universe, as parts of, or extensions to, their own self; an unshakable conviction of their rightness and virtue; and a stunted empathy for others. Such traits may find expression in pro- or antisocial ways.

An important difference between charismatic personalities and ordinary folk is that most of us attempt to fulfil our ambitions in realistic ways; that is, we take account of the needs and feelings of others. Normal people accept their limitations, their flawed but 'near enough' approximations of success in their attempts to live up to their ideals. For most of us our ideals and values are mere direction-setting standards that we try to live up to. We feel good when we measure up and bad when we fall short. Empathy with others shows us that no one is perfect, and this prevents the development of a sense of moral superiority. Hence, no unrealistic feeling develops that we are perfect while others are corrupt. The charismatic, however, lacks empathy. He identifies totally with his ideals and no longer measures his behaviour against them. He and his God (or 'ultimate concerns') are one, and there can be no half measures, hence he must always be right (Kohut, 1976).

Such persons pay a price. Their relationships are shallow because they have a double standard of reality; they relate with genuine concern to others at the same time as they see them as objects to manipulate. Further, the narcissist's ego-reflecting world-view is always under threat because people behave differently from how he wills them to behave. Although this difference can be excused — one's stomach sometimes behaves differently from how one wishes it would — nevertheless the logical implication is that the leader is not really in control, that there is an objective reality beyond his self. But to admit this is to admit the original traumatic circumstance that inspired the flight into narcissism. This must be avoided at all cost.

What the charismatic personality most lacks is a sense of his shared humanity with other people. He may accurately diagnose their problems and brilliantly solve them. He may even genuinely love the followers, as much as he is able, loving them quite literally as he loves himself. Yet they remain unreal to him because he must not acknowledge what it means to be a fellow sufferer, to feel alone and to have to adjust to an indifferent world, to have to reach out in need, trust and fear to another for help. He may have actually been alone and had to adjust, but he is rigidly fortified against the meanings of such events. They occur to him as strange inexplicable interludes in a continuum of mastery and dominance, self-sufficiency and control; he is 'phobic' about recognising any emotional vulnerability. Any outright opposition is countered with vociferous energy — what Kohut calls 'narcissistic rage' — a rage that shows by its extremity and persistence that he is more deeply wounded by affronts to his world-view than by any physical injury (Kohut, 1972).

The charismatic personality is fondest of the true believers who enthusiastically mirror his ego. Those who do not are resented. Despite the leader's wisdom and all-embracing facade, his acceptance of others exists only so long as his own needs are being fulfilled. When others behave contrary to his wishes he may respond with incomprehension or paranoia. For what he really empathises with most are shades of himself, and he attracts only those who are in tune with him. He is unable to empathise with people who are indifferent to him, whose needs do not mesh with his own. His inability to experience himself as vulnerable is like a chasm between himself and others, for vulnerability is a vital part of human reality; we are not gods or superheroes. Anyone who cannot experience vulnerability remains fundamentally out of rapport with ordinary people, no matter how successful their manipulations, or how sublime their 'truth' may appear to be. Because of this deficiency, the charismatic leader is not a great man; rather, he is a great actor playing the role of a great man, although if he plays his role very well he may indeed become great. Similarly, he cannot be his real, authentic self, but must forever play out his grandiose fantasy, although again, if he believes in this strongly enough it may become real for him, and anyway, it is probably the only reality he knows.

Hence charisma involves something of a gamble. It is a high-risk strategy for both leaders and followers, one in which the narcissist attempts to steer a path through life by posing as a leader and recruiting others as supports; others who are sufficiently needy to risk elevating him to leadership. In Churchill's case this required massive energy, intelligence and creativity as well as the daily anaesthesia of large amounts of alcohol, supportive others and a lot of good luck. Most charismatic personalities are not so lucky and many end in failure, although this may also have something to do with the magnitude of the tasks they set themselves, for they are most able to recruit a following among those facing major problems. Churchill and other leaders of his type seem able to transform their pathology into a kind of 'cosmic narcissism' that is basically prosocial (this will be discussed below). Cosmic narcissism still involves grandiosity but there is also a shifting aside of the ego, or at least of some functions of it; Kohut speaks of a 'well-integrated narcissistic self' (Kohut, 1966, p. 256). Hence the personality becomes a mosaic of regressive, realistic, and even transcendent features. In cosmic narcissism, while the external world and other people continue to be felt as extensions of one's self, the charismatic personality identifies with them so intensely that he equates their welfare with his own, even to the degree that his own physical survival is counted as being of less importance than their total welfare. Thus, Churchill would rather have died himself than surrender his nation to the Nazis. This is probably the most growth that can be achieved by highly narcissistic individuals.

Finally, Otto Kernberg's view of narcissism (Kernberg, 1975) is worth also considering here because he summarises 'the views of most streams of analytic thinking on the subject' (Symington, 2002, p. 58). Kernberg may actually be a little clearer than Kohut on the dual developmental origin of narcissism. For it can be objected that Kohut's two scenarios really constitute a bet both ways; if the child experiences a traumatic failure of rapport then that can lead to narcissism, and if he does not then that too can lead to narcissism. Such criticism may be unfair, but Kernberg seems more specifically to allow for a scenario wherein a developing infant may experience both a traumatic deprivation and maternal idealisation in parallel. Such a dual developmental pathway is by no means unknown for some individuals (Millon et al.,

2004), and there is even some research indicating that it may underlie pathological adult narcissism (Ramsey, Watson, Biderman, & Reeves, 1996; Stone, 1993).

This dualism of original experience may be related to the peculiar dualism of the adult charismatic personality. Personalities are usually described in terms of dichotomies; to describe someone as savvy is to imply that they are not naïve, while to say someone else is outgoing is to suggest that they are not shy. But a more sophisticated understanding reveals that such either/or conformations are misleading. We usually possess both sides of a trait but we use them in different situations. The woman who is quiet and shy when meeting strangers may be very gossipy among her friends, while the boss who appears dominant to his subordinates may be intimidated by his wife. People also change over time, and the polarities of a trait may be unequally developed at any particular moment or in any given context, but they are usually present. However, the charismatic personality is usually described as possessing a 'larger than life quality', or it is said that 'opposites abound' in them. (This characteristic is not confined to charismatic personalities of course; Oscar Schindler was both corrupt and virtuous at the same time.) It is as if the charismatic personality has both polarities of a trait developed to the extreme, and mostly quite independent of context. Common sense suggests that this probably originates somehow in childhood.

Kernberg identifies 19 characteristics of the narcissistic condition (see Table 1; Kernberg, 1975). These can be seen as symptoms of narcissism that may also be present in individual charismatic personalities, and as the same phenomena that Kohut has described but seen from a slightly different perspective. Sometimes only the words are different, but that may still be enough to clarify understanding. Hence, Kernberg's symptomatology is briefly presented at Appendix A in order to supplement the above discussion.[6]

Endnotes

1   I rely on Ted Hughes (1997) translation of Ovid's poem, and commentaries by Holmes (2001) and Ride (2002) in this account.

Table 1

Kernberg's Symptomatology of Narcissistic Personalities

- An unusual degree of self-reference in their interactions with other people.
- An inflated self-concept.
- Extreme self-centredness.
- Idealisation of people from whom they expect narcissistic supplies (that is, applause and tokens of affection).
- Arrogance, being a defence against paranoid traits.
- A shallow emotional life.
- Little enjoyment from life itself.
- Boredom and restlessness.
- 'Emptiness behind the glitter.'
- Little empathy for the feelings of others.
- Control of others.
- Very dependent on others for tribute and praise, but unable really to depend on anyone because of a deep distrust.
- Strong conscious feelings of inferiority and insecurity (in a functional charismatic personality such feelings would be unconscious or strongly denied).
- A great need to be loved by others.
- Coldness and ruthlessness behind the charm.
- Envy of others.
- Superficially there is a lack of attachments to others, but actually their interactions reflect very primitive and intense (attachments) of a frightening kind.
- Relations with people are exploitative and parasitic.
- Extreme contradictions, such as fluctuating inferiority versus grandiosity, pursuit of power versus impotence.

Source: Kernberg (1975)

2   Ovid was not alone in his analysis of narcissism. Shakespeare in his Sonnet 62 begins: 'Sin of self-love possesseth all mine eye, and all my soul, and all my every part; And for this sin there is no remedy, it is so grounded inward in my heart'.

3   Subsequent psychoanalytic theorists have challenged Freud's early formulation. Michael Balint (1968) and Ronald Fairbairn (1952) have argued that we are able to relate to others, to discriminate them from ourselves, right from the start of life, and the research of Stern (1985) on infant behaviour appears to support this. More recently Symington (1993, p. 120) has proclaimed that 'the only narcissism that exists … is secondary narcissism'. The difference seems to hinge upon whether one views nar-

cissism as a 'stage' of development as did Freud, one that begins and ends at certain points after which it gives way to more realistic relations with others; or whether, with Kohut, one prefers to think of it as a developmental stream that is ever-present throughout life as a precondition for healthy development at all stages of growth, including old age in which the now-transformed narcissistic stream still has work to do in the generation of wisdom and acceptance of death. Not being an analytic theorist I have no opinion on this debate and would be happy to construct a theory of charisma from either perspective, but it does seem that both positions require more development before one could commit to it over the other.

4 Kernberg (1975, p. 229) also has noted that, 'What distinguishes many of the patients with narcissistic personalities from the usual borderline patient is their relatively good social functioning, their better impulse control ...'

5 Kohut does not go deeply into quite how this happens, but we can make some informed guesses. Personality is largely socially constructed, but if one regards the social world as merely an extension of one's self, that is, as a part of oneself, then major areas of psychic functioning change their meanings drastically; for example, defence mechanisms. What would such a person have to defend himself against? Himself? We need not dwell on this point except to observe in passing that seeing the world as a part of oneself changes utterly the internal relations of the psyche and the meanings of such threatening feelings as anxiety, which probably become split off or greatly distorted. As such it may bring a powerful fearlessness to one's experience of the external world.

6 Interested readers may peruse Summers (1994) for a fuller summary of both Kohut's and Kernberg's approaches, or consider Benjamin (1996) for yet another perspective using the same theoretical tools.

CHAPTER FOUR

• • • • •

The Charismatic Intellectual:
Sigmund Freud

He who has been the undisputed darling of his mother retains throughout life that victorious feeling, that confidence in ultimate success which not seldom brings actual success with it.

Sigmund Freud

Although the popular mythology of charisma has it as an attitude of the needy, it may be that charisma is more characteristic of intelligent and sophisticated people than of the superstitious and more easily manipulated masses, who may be too downtrodden to dream of salvation or to idealise a hero. It is relatively easy to appear charismatic to needy or uneducated people; one of the leaders in my original study (entitled *Prophetic Charisma*) told me that when he was starting out as a young pastor he was told by his supervisor, 'In difficult times anyone who has a secure marriage, a good job and a happy family, is looked to for leadership'. But the needs of simple people are likewise simple; their interests are more focused on material wellbeing, and perhaps their fear of the unknown is more severe and thus they are less inclined towards heroic risk-taking. In *Heart of Darkness*, the definitive novel of charisma, Joseph Conrad (1902) has Marlow say disparagingly, 'A fool is always safe', by which he meant that one who is unaware of their depths is immune to charisma. This might apply more to less-aware folk than to the better educated.

In contrast, many educated people imagine that they are immune to the allure of charisma, as if they are in some way above it, insulated from its seductive charm by their intelligence and knowledge. Something about this assumption jars, for the most highly intellectual endeavours seem to be especially prone to charisma. As the philosopher Michael Polanyi has shown, the history of big science has largely been a history of powerfully forceful men with overweening ambitions and giant egos (Polanyi, 1964). One need only think of personalities such as Beethoven and Picasso to suspect that perhaps the art world may be similar. Even some of the highest academic institutions have, according to some accounts, long traditions of charismatic gurus setting the tone — F.R. Leavis and Ludwig Wittgenstein at Cambridge, and Maurice Bowra at Oxford to give 20th century examples (Malcolm, 2008, 2009). The intense scrutiny of the intellectual world is very demanding, and perhaps this drives charismatic processes higher. There are several major analyses of the great scientific paradigm shifts of history that have shown fairly conclusively that these changes occur not solely for logical or scientific reasons, but are in fact driven largely by social factors, sometimes including propagandising by elite intellectuals for their hobbyhorse theories (Feyerabend, 1975; Kuhn, 1970). To push an unfamiliar idea through intense intellectual resistance may take much more than mere facts and reason.

Hence, to consider just the leading thinkers cited herein, several were highly charismatic themselves (perhaps it takes one to know one). For example, Heinz Kohut was initially rejected for training as a psychoanalyst, probably because of his own narcissism. However, he turned this rejection into a triumph. The board that rejected him said that he was unfit to train as an analyst and recommended that he undertake further personal psychoanalysis. He did so, but he cannily selected as his analyst someone who was also licensed to conduct training analyses and who carried much weight inside the institute (in psychoanalytic training one must undergo a 'training analysis' as distinct from any personal analysis one may have had). When the time came for another application this analyst might be counted upon to advocate for him and to credit his personal analysis as a training analysis. Sure enough, he was accepted for training next time around, his personal analysis was credited as a training analysis, and he eventually became a fully qualified analyst. Many years later he was seen as a psychoanalytic 'New Messiah' and his 'disciples' dubbed themselves 'Kohutians' and were viewed as 'cultish' (Malcolm, 1980; Strozier, 2001).

It is tempting — perhaps compelling — to see charismatic influence at work here. This story also ought to be a salutary lesson to those who imagine that procedures such as questionnaire screening might weed out narcissistic personalities from the ranks of therapy, management or the clergy (such suggestions are made from time to time). Charismatic individuals could easily circumvent such obstacles to get where they want.

Similarly, Max Weber was himself highly charismatic, as his acolyte Karl Loewenstein has described.

> There under the rose arbor, at a table heaped with books, sat Max Weber ... What followed in the course of this first encounter with him was a wonderful and, for me, decisive revelation. For he began to block out a sociology of music for my benefit ... I had thought that music flows from emotional and aesthetic sources and I now drank in his words as he explained that music too has rational and sociological foundations ... I must have sat with him for more than two hours ... When at last I took my leave and went out on the road into one of those glorious Heidelberg June evenings I was literally drunk. I was at a turning point in my life. From that moment on I had taken the oath of fealty to him. I had become his vassal ...
>
> [On Weber's return to the podium] It was one of the most magnificent addresses I have ever been privileged to hear ... he spoke extemporaneously, merely using this brief outline and for more than two hours he kept the whole audience spellbound ...
>
> [Later in his diary Loewenstein wrote] When he ponders his face contracts like the sky before a thunderstorm. It is a manly face; something elemental, actually titanic, emanates from him. He speaks freely in his resonant voice, using a magnificently controlled German, every word in its

> proper relation to the context … It was more of a political sermon than a learned discourse, coming straight from a great overflowing heart and sustained by a breadth of knowledge and thought that again and again gave something to us by the very context in which it all was placed … For two hours a sold-out house listened in breathless suspense. Those who have not heard him speak cannot imagine the power and expressiveness of his language …

> Max Weber was a daemonic personality. Even in routine matters there was something incalculable, explosive about him … he was utterly fearless … he was also the charismatic man that he himself described. He had that exceptional gift of casting his spell upon everyone he [His wife] sent me a picture of him on his deathbed. It has hung above my desk for more than forty years … I have never met another who could compare or in any way come up to Max Weber. (Loewenstein, 1966, pp. 93–103)

One of the best descriptions of a charismatic intellectual concerns the pioneering psychologist E.B. Titchener (1867–1927), as reported by the historian of psychology Edwin G. Boring. Once a highly influential psychologist but now merely a footnote in introductory texts, Titchener is almost the paradigmatic charismatic scientist. Very little of his work remains, yet if one digs into his writings one finds that his thoughts are clear and his ideas are persuasive. It is just that ultimately they were not supported by scientific research. He is credited with introducing the 'new psychology', the experimental psychology of Wilhelm Wundt and others, to the United States, thus moving psychological research away from 'mental philosophy' to empirical psychology as it is understood today. Through systematic introspection he attempted to develop a theory of consciousness, but the very research methodology that he championed gave his ideas little support. Eventually, he dropped out of the American psychological establishment altogether and formed his own school known as the Experimentalists. Despite his lack of success he never abandoned his introspective 'structuralist' approach. Boring has recounted his experience as a young researcher studying under Titchener at Cornell University.

Psychology at Cornell — at least the orthodox psychology that centred in the laboratory — revolved around and was almost bounded by the personality of E.B. Titchener ... And what a man Titchener was! He always seemed to me the nearest approach to genius of anyone with whom I have been closely associated. I used to watch my conversations with him hoping to gain an insight into why his thought was so much better than mine ... He expected ... deference and ... loyalty from the juniors of his immediate circle ... Those who broke the code of manners he ostracized at least temporarily and often permanently. It took over a year before Titchener's loyal adjutant H.P. Weld, was received back into the communion ...

One may ask ... how it came about that Titchener could dominate and control the conduct of his in-group. He did it by the magic of his personality, by his social charm, by his identification of himself with all those others, mostly his juniors within his circle. You decided eventually from his brilliance, from his erudition, and his long list of honours that he was a Great Man ...

After Titchener's death I went over the 212 letters I had had from him since leaving Ithaca to see what there was in them worth publishing; there was a little but not much. All the rest was this magic personalized glue that held the in-group together, was Titchener's interest in my personal affairs, or his plan that 'we' should promote an appointment that would add another laboratory to 'our' group of loyal institutions. (Boring, 1961, p. 20)

In a letter to the psychologist Robert Yerkes in 1945, Boring expressed this point more forcefully:

Titchener's letters! You know I had a treasured folder of 212 of Titchener's letters and felt that something important must be done about them ... Then I sat down to them and went through them, went through them with a great sense of depression and disappointment. What seemed so great and important and marvellous at the time turned out to be small and petty and personal ... It was a very disappointing experience, but also a maturing experience. It showed of what stuff hero worship is made. (Boring, 1961, pp. 110–111)

Boring continued his 1961 account:

> Freud required loyalty from his in-group and held it together in a similar way. With [William] James the in-group was formed through the democratic charm of friendship, but with Titchener and Freud the friendships were authoritarian and personal. Both men needed to play the father role, dispensing much real kindness to those disciples who avoided transgression. I never broke with Titchener, but neither did I follow him blindly. On theoretical issues I differed vehemently in the privacy of the laboratory and recognized the importance of Titchener's views by calling myself a 'heretic'. Others seemed to me to accept unquestioningly the pronouncements of this oracle and often I found myself arguing hotly against deference to authority in matters of fact or logic ...
>
> Titchener had far more influence upon me than any other person in my professional life, the brilliant, erudite, magnetic, charming Titchener, who interested himself in your research and writing and how your wife cooked the mushrooms, helping in big and little things, but demanding loyalty, deference, and adherence to the advice so freely offered. In the last twenty years of his life Titchener had few contemporary friends and remained surrounded by younger disciples. Was I one? I used to call myself a 'heretic,' being more aware of my dissents from Titchener than of my agreements ... (he) remained always a fascinating subject ... On 3 August 1927 the impossible happened. Titchener died. It was as if the Ten Commandments had suddenly crumbled. (Boring, 1961, pp. 21–26)

It is important to appreciate how the language used by believers to describe charismatic personalities correlates with the levels of secular sophistication each possesses. Jesus is God, Swami Vivekananda 'transcends genius', Hitler is 'a genius' and Titchener 'approached genius' (Oakes, 1997, p. 186). The language used may differ but the emotions and the relationships are very similar. Note also the details of the descriptions; Loewenstein was at a turning point in life and Boring was a student; the adjectives used include 'incalculable ...

daemonic ... magic ... fascinating ... [and] magnetic'; the relationship is one of 'fealty ... [like a] vassal ... [or a] disciple', and it feels like being 'literally drunk'; the leader's teaching is a 'revelation', and when he dies it is 'as if the Ten Commandments had suddenly crumbled', and a cherished last picture remains above a desk for forty years. These are frankly religious sentiments.

But without doubt the most revealing example of the influence of charisma on the intellectual world, and thus on culture and everyday life, concerns Sigmund Freud, who could arguably be said to have had a greater influence on the modern world than any other thinker.

Freud began his momentous project in Vienna in the 1880s as a young doctor with an interest in mental illness. He apprenticed himself to several famous physicians and extrapolated from their work, but when he found he needed to go beyond these figures he undertook a remarkable experiment. He turned his psychological tools on himself and his own life history in a singular effort at self-analysis. He was not totally unique in this endeavour, Marcel Proust was also to undertake something similar, but Freud was working self-consciously within a scientific framework, and attempting at the same time to develop a theory that might be applied to everyone. He also had the advantage of having previously analysed the lives of many disturbed patients, and of having picked the brains of some of the best psychiatrists of his time. The result bore fruit as *The Interpretation of Dreams*, the 1900 magnum opus still considered his best work. He later went on to found psychoanalysis, the movement that dominated psychiatry for about 50 years. His collected works are vast, and many of his key concepts such as the Ego, Superego and the Id, the Oedipus Complex, Anal Retention and Penis Envy, have infiltrated all levels of culture. However psychoanalysis has recently gone into decline, partly because its claims have not been verified, but mostly because increasingly it has come to be seen as something of a 'scientific fairy tale', and Freud himself as a somewhat dubious figure.

A huge research effort into Freud and psychoanalysis has been underway for the last 40 years, and a consensus has now emerged that while Freud was a great thinker, he was a less-than-great human being. Although his ideas are immensely provocative and fertile, he was also something of a 'ruthless, devious charlatan' (Clare, 1997, p. 10). There are several excellent biographies and criticisms available,[1] but this chapter will focus on the charismatic elements in his life.

Sigmund Freud's childhood was traumatic and complex. His father Jacob had been married previously (possibly twice), and his wife (or wives) had died, leaving him with two sons whom he took into his wool-merchant business. Jacob then married Amelia, 20 years his junior, and she gave birth to eight children in ten years, of whom Sigmund was the first-born; his two half-brothers were aged 16 and 20 when Sigmund was born and did not live with them. At first the family lived in a one-room apartment in Freiberg. Then Jacob's business failed — he was never again able to earn sufficient money for his family — after which they moved first to Leipzig and then to Vienna; Sigmund was aged three and four at the times of these moves.

During Sigmund's childhood there occurred a series of bereavements, beginning with his immediately younger brother Julius, who died when Sigmund was almost two. This probably left his mother depressed, at least for some time. She was also hospitalised or ill from time to time due to bouts of tuberculosis. These maternal absences probably produced the dream he had when he was aged nine (recounted in *The Interpretation of Dreams*) in which he saw his mother dead and awoke terrified. Other losses included his paternal grandfather, who died six months before his birth but whose death no doubt cast a shadow over his father and several other close relatives, and his beloved nursemaid after she was caught stealing and sent to prison; he was three at the time. Other shadows included Jacob's loss of his first two wives and the death of Amelia's brother. In sum, Sigmund was born into a family atmosphere of mourning.

Some have pointed to this series of stresses and losses as the main traumatic ingredients in Freud's early life. While these were obviously influential, such an approach tends to overlook the sheer intensity of the 'normal' occurrence of eight children born in the space of ten years, and the resulting competition among them for limited parental resources. Sigmund would have experienced very little 'quality time' with either parent (Amelia worked in Jacob's business), maybe none at all. Combine this with the atmosphere of mourning; the business failure of his father, who seems to have been a genial but ineffectual man, and the resultant poverty and dislocation caused by this; the one-room accommodation; his mother's tuberculosis; and then add to all of this the sense of isolation from being new arrivals in both Leipzig and Vienna, as well as being members of the Jewish minority, and it becomes surprising that Sigmund did so well to adjust and ultimately succeed.

There were other peculiarities. Sigmund was 'mothered' in some sense by three women in his early years — his mother, his nursemaid and his half-brother Emmanuel's wife, Marie. In addition, his father's ineffectualness was rescued by Amelia's dominance within the family; she seems to have been a woman of strong will and convictions, and the family was a matriarchy. Finally, there is the fact that as the first-born child, the oldest and for nine years the only son, Sigmund would have not only carried the hopes for the future of the family, but also have been seen by his parents as in some way compensating for the loss of the second son and the arrival of six daughters. As the quote that heads this chapter implies, he really was his mother's (and father's) darling. He was born in a caul, which in the folklore of the day marked him as special, and his mother called him 'mien goldener Sigi' (my golden Sigmund). He was clearly the privileged favourite (he had the only oil lamp in the family — everyone else had candles), even becoming something of a tyrant within the household; when a sister's piano practice disturbed his study he issued an ultimatum to his mother and the sister was forced to abandon her piano lessons.[2]

So many pathological strands could be inferred from this material as to keep a team of psychologists busy for months. To consider just the quote that heads this chapter, we might ask why he overlooked his father, or why his focus was on success rather than happiness. In the light of his subsequent theorising we might also question his use of the word 'darling', which has erotic associations. Knowing the reality of Freud's childhood we might also wonder what the quote was designed to disguise or deny. In fact, much of Freud's later thought can be traced back to this childhood nexus. The conflict model of psychoanalysis in which emotional, sexual, familial and social forces are in conflict, and wherein they compete for expression and dominance, clearly came from his childhood experiences. His Oedipal theory in which a son usurps his father, his reductive anthologising wherein he claimed to detect morbid and perverse motives and murder and incest themes 'beneath' the warmest demeanours of quite decent people, his analysis of love as merely a modified survival instinct ('aim-inhibited lust' he called it), and his need for enemies and his suspicion of dissenters, all combine to give us a picture of how he experienced life as a child. The Freudian universe was a cold and dangerous place.

The eventual result was a man who experienced a powerful hunger for achievement, and such a great need for fame that he was prepared to act ruthlessly to gain it. He was personally distant and suspicious of intimacy and even somewhat paranoid, and he found solace from the world in his books and theorising. His tremendous intellect blossomed as his parents guided and fed it in the hope of compensating for their own disappointments. Unlike Churchill, Freud did not need to spin a fantasy of compensatory grandiosity; this was provided for him by his parents (it was virtually mandatory). Despite his privations he was subjected to 'baby worship' from his earliest childhood, given the message that he was special, a sacred child, until this became the only kind of relationship he could feel comfortable in. He concluded (or was overtly told) that he was destined for great things, and implicitly, that with the attainment of his greatness the healing balm of love would flow freely in his family.

Much of Freud's behaviour as a young man was typical of many struggling and gifted young intellectuals of his time. He was a brilliant anatomist and diagnostician and he soon reaped professional respect for this, publishing learned papers on neuroanatomy and diagnosis, and lecturing visiting physicians in 'a sort of pidgin-English'. This had a significant psychological effect on him, changing his personality and marking him — in his estimation — as one destined for greatness. Ellenberger describes the effect as similar to that undergone by Robert Bunsun after he discovered spectral analysis. Apparently, after this success Bunsun changed his entire demeanour and from then on 'he bore himself as a king travelling incognito'

(Ellenberger, 1970, p. 449). Erik Erikson's daughter, in her book *In the Shadow of Fame* (2005), has also described how, after the great success of his book *Childhood and Society*, her father 'began to take on a new social persona ... his new confidence in the power and importance of his ideas greatly enhanced his aura' (Bloland, 2005, p. 93) despite previously being uncomfortable and restless socially. Freud too, took immense — perhaps grandiose — pride in his achievements.

However, this early success, while it confirmed Freud's sense of a special destiny, was not sufficient to rest upon, nor to sustain him. He continued to explore (his innate creativity and curiosity would have guaranteed this), and to apprentice himself to senior scholars. This is also typical of charismatic personalities, many of whom are themselves prone to charismatic attachments to others who ultimately disappoint them; a leader in *Prophetic Charisma* told me that the reason he set up his own church was because he could not find a teacher that satisfied him, although he had tried many. They seem to be genuine seekers who, while needing fame and a following, in their early years nevertheless may attempt to find a realistic resolution for their issues in the inspired guidance of a wise elder (or elders) who ultimately fails them.

Freud had several teachers and mentors with whom he became infatuated. What is striking about these early relationships is that he lived at a time of fertile speculation within psychology and medicine, and that most of these mentors entertained crackpot theories of their own, despite their academic prominence. It was a cusp time in intellectual history, when several disciplines including psychiatry and psychology were coalescing into something like the shapes they have today. Psychology had recently departed the philosophy faculty — the first experimental psychology laboratory was begun by Wilhelm Wundt in Leipzig in 1875 — and 'scientific psychology' (as opposed to 'mental philosophy') was something of a Holy Grail for many; Freud even wrote an unpublished book on the subject.

For example, Gustav Fechner, discoverer of a fundamental psychological principle known as the Weber-Fechner Law that made him famous within the ranks of psychologists (Freud was to call him 'the great Fechner'), published pseudonymously a book titled

Comparative Anatomy of the Angels in which he extrapolated from evolution in the animal kingdom and from amoeba to man, to what he concluded would be the ideal form of the highest earthly beings — angels. He concluded that they would be spherical and communicate by a language of luminous signs.

Similarly Ernst Haeckel, formulator of the 'Bioenergetic Law' that underlies several modern psychological theories as diverse as Jean Piaget's theory of child development and Rudolf Steiner's theory of education, also founded the new pseudoscientific religion of 'monism', that he hoped would rid Germany of Christianity and replace it with a glorified modern science (Adolf Hitler's philosophy of history was strongly influenced by Haeckel's theories).

Another German evolutionary thinker Wilhelm Bolsche, whom Freud cited in his *Introductory Lectures on Psychoanalysis*, theorised that simple organisms used the mouth as their reproductive organ and thus sexual reproduction was originally 'a sort of higher eating'. By implication, the sexual consciousness of the human child progressed from a stage of perverse animal sexuality through that of the fish and reptile and ultimately to that of the adult human. Bolsche's ideas profoundly influenced Freud's main mentor Wilhelm Fliess.

Other theorists — including several of Freud's own teachers such as Ernst Brucke, Theodor Meynart, and Sigmund Exner — promoted or engaged in what has been called 'brain mythology', producing 'vast speculative structures scarcely less extensive and bizarre than the psychomechanical model elaborated by Freud' (Webster, 1996, p. 178). Jean Charcot, an eminent authority with whom Freud studied in Paris, pioneered hypnosis and championed the diagnosis of hysteria and then claimed to detect it everywhere, freely applying it to all manner of unrelated ailments (it turns out that there is no such illness), and was perhaps the most bogus of them all. His work scandalised and embarrassed his colleagues at the La Salpêtrière hospital, and there remains the suspicion that at least some of his experiments were fraudulent. He once diagnosed a man as suffering from syphilis and then ordered that he be hung from his neck to relieve his pain by 'lengthening the nerves' (the treatment hastened the man's death).

Freud remained in thrall to these dubious fellows for the rest of his life, even naming his first-born son Jean Martin after Charcot. Most of all he emulated their theorising, taking on their intellectual strategies and finally concentrating them into a single unified discipline that became psychoanalysis. He was not alone in this, however; for within European intellectual circles it was a time when speculative theorising was rampant; when even great scholars made appalling mistakes in their pursuit of some fundamental discovery in the new science of psychology that might 'open all secrets with a single key' (Webster, 1996, p. 196). Franz Josef Gall was such a figure; a brilliant neuroanatomist who made many fundamental discoveries that still underpin modern medicine (his early achievements far outshone those of Freud) but who then went on to create phrenology.

What is perhaps most significant in all this is that like all charismatic personalities Freud was attracted to a frontier space with ambiguous parameters, where he could pose as the bearer of a great new solution to intractable problems, and where he could gain the mastery and adulation of others that he felt was his birthright. However, he made several false starts and he behaved shabbily right from the beginning of his career.

For example, as a young doctor he was frequently depressed and he took cocaine for this (quite legally), soon becoming an enthusiast and proselytising the drug to others. However, his irresponsible early experiments with cocaine led to his friend Ernst Fleischl becoming addicted. Fleischl had major problems before this, which Freud's treatment worsened, but he nevertheless wrote up a case study of his friend as a successful treatment.

Later Freud co-authored with Josef Breuer the book *Studies in Hysteria* which claimed (among other questionable matters) a successful treatment through analytic catharsis of Breuer's patient Anna O. Detective work by Henri Ellenberger almost a century later showed that this woman emerged from Breuer's treatment far from cured (she entered an asylum shortly after) and indeed, the treatment did not even proceed as Freud and Breuer had reported (Ellenberger, 1970). Freud must have known this case study was highly inaccurate, but he published it anyway, only falling out with Breuer much later over quite a different matter. Breuer had been immensely generous to Freud, helping him into his practice and advocating for him to colleagues, but Freud could not tolerate it when Breuer did not fully support his ideas, and later he simply cut him cold in the street (Breger, 2000). It seems that Freud was never able to accept criticism or dissent (Webster, 1996). Later he bribed his way into professor status at the university, such behaviour being quite common in Vienna at that time (Webster, 1996).

What was perhaps Freud's most egregious misbehaviour came later in his career when a 14-year-old girl was referred to him for stomach pains. He diagnosed 'hysteria' as usual and claimed to have cured it, discharging the child, who died of undiagnosed stomach cancer some weeks later. In subsequent writing about this case he nevertheless continued to claim it as a 'cure', which had unfortunately disguised another quite independent complaint. He even wrote of this case many years later in a disarming way, with regret, as if to confess a mistake due to inexperience, though still claiming it as a successful treatment.

In later years, another of his dubious behaviours was to spread a myth of his great struggle against ridicule and ignorance in the name of truth. He inflated the scepticism and the caution that he and his ideas faced in his early years, so as to make it appear that he had battled heroically against bigoted intolerance and prejudice. But as Frank Sulloway (1979) has shown, Freud never received the kind of opposition he claimed, and was moderately well received, or at least listened to seriously, by most audiences. When he returned from Paris enthused by Charcot's ideas he gave a lecture on them to the Viennese Society of Physicians, that he claimed was greeted by cold incomprehension. In fact his talk was met with only cautious reserve, but he was hoping for the same adulation in Vienna that Charcot had received in Paris. The modest reception he was given bitterly disappointed him, and it still rankled with him 40 years later when he recalled it, casting himself as the genius-innovator facing the narrow-minded, resistant, ultraconservative authorities defending their status-quo.

This points to another major theme in the lives of charismatic personalities, their enormous sense of entitlement. Freud 'knew' he was a genius from very early in his adulthood (probably even much earlier), and consciously crafted an image of himself as a heroic genius. As early as 1885 when he was 28, he deliberately destroyed all his notes, letters, scientific excerpts and manuscripts in order to create his own legend. Writing to his fiancée Martha about this, he asserted that he would not make it easy for future biographers to slot him into their pet theories about 'The Development of the Hero' (he had done nothing heroic up to this point). He carried out other purges in 1907 and again toward the end of his life. He also tried to suppress his correspondence with Wilhelm Fliess, which fortunately survived, because he knew it would compromise his myth.

Freud's grandiose sense of entitlement probably derived, in part, from the influence of his parents and their expectations that he would become great. But it may also have been fed by the privations of his childhood; having suffered much he sincerely believed he deserved much. He even wrote a paper to this effect about 'the exceptions', those individuals who do not feel the same guilt and obligation to moral rules that others do, because they believe that some suffering they have undergone excuses them from such human responses and obligations.[3] Continuing his childhood sense of deprivation into adulthood, he experienced the world as not giving him his due. No matter how much he received it was not enough; he had a 'surplus trauma'.

At other times Freud distorted his case studies to present his findings in the best light. First he claimed to find evidence in his 'research' for his patients' hysteria being caused by childhood seduction, then it became incest, and still later this was modified to fantasies of seduction and/or incest. At each point he bent his material to the hypothesis he favoured, while downplaying his earlier positions. Perhaps his most infamous case involved his greatly exaggerated claim of success in his celebrated case of the Rat Man (Breger 2000, p. 180). Then there is the equally infamous episode of his bungling friend Fliess's nose operation on Emma Eckstein, which Freud covered up. Through all of this, Freud showed himself to be ruthlessly self-promoting and ultimately indifferent to the sufferings of others where these conflicted with his own ambitions. When it suited him, he also slandered dissenters mercilessly, his disciple Sandor Ferenczi coming in for particular vitriol. He expelled several of the leading lights from the psychoanalytic movement for questioning his theories, and throughout his adult life he could only maintain friendly relations with those who supported him uncritically.

Freud seems to have been fundamentally dishonest, but in a very subtle and calculating way. When devising his theories he adopted the 'rhetoric of empiricism' and consistently presented his discoveries as if they were the result of numerous 'laborious and detailed investigations', when in fact he often extrapolated confidently from just a case or two (Webster, 1996, p. 206). This is instructive; while the simplest interpretation is merely that he lied, as indeed he literally did, a more sensitive interpretation would be that he lied in the sincere belief that he was a genius who possessed such superior acumen that he could legitimately extrapolate from a single case, whereas other lesser beings would need many such cases to reach the same conclusions (he was an 'exception' again).

This may have been partly because he felt that he had gained such license with his self-analysis. In 1897, with the support of Wilhelm Fliess, he had embarked upon this career-defining exercise. He used the intellectual toolkit he had assembled, including the many contributions he had taken up from others, to analyse his own life and experience. He hoped to illuminate the structure of the unconscious mind (an idea that went back at least to St Augustine and Plotinus), as well as the psychic life and development of children. Along the way he 'discovered' infantile sexuality, repression and the Oedipus complex. This exercise was even accompanied, during the years 1894–1899, by the classic 'creative illness' that mystics and artists sometimes experience (Ellenberger, 1970).

Leaning on Fliess for support throughout, Freud's self-analysis was a remarkable intellectual adventure that provided him with insights he would use for the rest of his life. The great virtue of this exercise was that by analysing his own life he had the best chance of knowing for sure (or as near as he could) what the objective facts were. Then, by using the psychological tools he now possessed, he could relate this material to his dreams, thoughts, behaviours and feelings, and thus he could come to understand himself much better and hopefully develop a technique for analysing the lives of others. Such a technique might be of great benefit to mankind, perhaps even curing serious pathology. Hence, Freud's project was rooted in the search for truth, undertaken in the spirit (if not the actual practice) of scientific inquiry, invested with immense hope, sincerity, compassion and altruism, and undertaken no doubt with humility, courage and rigour, by a passionately idealistic young man of genius. This proved a heady brew; at one point he became so grandiose and enraptured with his 'discoveries' that he boasted to Fliess that he could cure every case of hysteria (Farber, 1961). Unfortunately, he got it all terribly wrong.

Freud was a product of his time and his family. This was, after all, the Victorian era when the morality of the day dictated sexual repression, ignorance and denial, producing an inevitable reactive salaciousness. Freud's marriage to Martha was initially quite sexual but soon became much less so, leaving him frustrated, a state that reinforced his reliance on self-control, intellectual detachment and inner containment, the survival strategies that had enabled him to endure his troubled childhood. Now, in his self-analysis, where he encountered love he tended to see sex, and where he encountered ambiguity he tended to assume tension and conflict, concocting his Oedipal fantasy of incestuous desires and murderous intentions.

Freud made the same mistake of many therapists and clients. He created a distracting yet valorising screen of colourful material to avoid his real issues. Most clients who come for therapy are ambivalent; they want help but are anxious about how much pain it may involve. Or, they may sincerely want change, but preferably to someone else rather than to oneself. People who attempt self-analysis are especially prone to this trap, deceiving themselves into believing a comforting myth. It seems reasonable to assume that one's madder thoughts and feelings signal some extraordinary and intense formative experience. Some clients, very early in therapy, blurt out a litany of painful, shameful or tragic events that they have experienced. These may include the deaths of loved ones, violence suffered or inflicted, sexual or emotional abuse and traumatic events that they have known or had done to them. It is easy to assume that these events are in some way related to their current distress, as indeed they might be. But often the therapist has to quieten such a client down and explain that not every bad experience results in trauma. Sometimes the really destructive and character-distorting influences in life come more from an indifferent emotional climate that allowed such abuse to occur, rather than from actual events in themselves, painful as these may be. Or they might derive from chronic emotional neglect, or conditional love, or cold, unfeeling parents.

In his self-analysis Freud was partly led by Fliess, who had some especially mad ideas, even for his time. Freud took the notion of infantile sexuality from Fliess, and also the linkage of unconscious contents with specific body parts, as well as other eccentricities. Consequently, he came to believe that the most exotic, colourful and extreme elements of his experience accounted for his personality and inner conflicts, fearlessly confronting what he imagined might be the worst about himself. These elements had some resonance with his actual experience, for he probably had eclipsed his father in his mother's affections, and with six siblings he probably desired more from his mother than he received, although love rather than sex, and he probably would have liked to have murdered various family members at times (as we all do). His Oedipal scenario had the effect of distracting his attention from a less romantic but infinitely more painful truth; that his childhood of loss, pressure to excel, sibling rivalries, the absence of an idealisable paternal role model and the lack of unconditional love, had distorted his basic humanity into the ruthlessly ambitious and depressive person that he had become. Rather than revealing this truth, his Oedipal romance enabled him to avoid it, while giving him a story that affirmed his self-image as one who contained the darkest that was humanly possible, but who had heroically transcended this darkness through self-control, intellectual detachment and inner containment, the survival strategies from his childhood, as well as courage, strength and intelligence (of course).

There followed his books *The Interpretation of Dreams* (1900), and then *The Psychopathology of Everyday Life* (1901), in which he applied the lessons he had learned from his self-analysis to the dreams and utterances he had collected from patients and friends, analysing the contents and exposing the hidden meanings, making sense of what at first seemed bizarre and meaningless. By implication, a similar sort of analysis could be undertaken with all manner of materials, including the utterances of therapy clients and the creations of artists. Now Freud had the fundamentals of psychoanalysis, and from these he not only created the first modern 'school' of psychotherapy, he also continued his speculative adventures — backed up by pseudo-scientific methodology — in several directions. He devised a theory of the mind that has been enormously influential despite having virtually no scientific support, and he applied his theory to art, biography, history, anthropology and religion, and made significant contributions to each field. Yet because of who he was and his time and place, Freud made some colossal mistakes. His essay on Leonardo da Vinci hinged on a mistranslation, and he saw sex in everything, at one time telling his protégé Carl Jung that they had to make an 'unshakeable bulwark' of his theory of sexuality.

Nevertheless, his work has formed the base upon which later generations of scholars have built a much more enduring structure. Further, that we can pick apart his mistakes and correct them is partly due to improvements made in his system by his intellectual descendents, especially the ego psychologists such as Karen Horney and Heinz Hartmann, the existential analysts Erich Fromm and Rollo May, and modern theorists such as Heinz Kohut, Jacques Lacan and Otto Kernberg, as well as analytically oriented critics such as Frederick Crews and Henri Ellenberger, who have progressively realigned psychoanalysis with common sense and the realities of daily life. (It is perhaps yet another testament to Freud that sometimes his critics have themselves made fertile mistakes as did Jeffrey Masson when, after making a major contribution with his publication of *The Complete Letters of Sigmund Freud to Wilhelm Fliess: 1887–1904* (1985), he also published *Assault on Truth: Freud's Suppression of the Seduction Theory* (1984), which forced scholars to re-examine Freud's early work in new ways.) Modern analysis is quite a different beast from that which Freud conceived. What really endures is Freud's creative attitude, a subtlety of questioning that at its best is both profound and infinitely fertile.

In sum, Freud created the most influential body of thought in modern times, and regardless of whether it is right or wrong it has been of monumental importance in several fields. Despite what modern scholars conclude about him — Peter Watson, author of the recent intellectual blockbuster *Ideas: A History from Fire to Freud*, concluded that 'Essentially he made the whole thing up' (Watson, 2005, p. 728) — it is hard to argue with success. And, as at least one contemporary theorist has argued, 'Freud was not always wrong' (Yalom, 2002, p. 217).

But one can judge success, evaluating how the successful person behaves in their hour of triumph and what kind of legacy they leave behind, and it is in this arena that Freud's charismatic behaviour came closest to that of a cult leader. The gatherings of the Vienna Psychoanalytic Society that he presided over were analogous to religious rituals, and his 'church', as it developed, became as secretive and mysterious and any Masonic order. At one point Freud had a special inner circle, his 'Secret Committee', issued with special rings, and he designated Carl Jung as his 'Crown Prince'. He fostered base toadyism, capriciously promoting and demoting followers who favoured or displeased him. Under the guise of rational scientific debate he argued down critics and defended his pet ideas energetically. He became the kind of tyrant in his movement that he had been in his family, controlled and reasonable on the surface but ruthlessly determined to have his way and, in time, this led inevitably to the defections of his brightest followers, including Alfred Adler, Carl Jung, Otto Rank and Sandor Ferenczi. As Thomas Szasz described it:

> Freud's leadership was deceitful. He created a pseudo-democratic, pseudo-scientific atmosphere but was careful to retain for himself the power to decide all important issues … Freud's essential concept of leadership seemed to be to bestow tokens of power on his competitors only to discredit them if they dared to use it. (Szasz, 1963, p. 153))

By misrepresenting his case studies, by expelling dissenters and by his authoritarianism, Freud showed that he had come a long way from where he had started out. Not, perhaps, so far ethically from the driven young man eager to make a name for himself at any cost, but certainly very far from the passionate idealist who hoped to cure mental illness. A change had gradually come over him. He seems to have become disillusioned, less generous and hopeful. Certainly World War I affected him greatly, as it did everyone, and it is likely that his pessimistic formulation of the Death Instinct derives from his ruminations on this, but it may also have had something to do with his state of mind at that time. The pain of his mouth cancer and the more than thirty operations he underwent for this no doubt wore him down (he apparently used cigar smoke to disguise the smell of his rotting flesh).

There were other disappointments too, more personal ones, especially the way his marriage to Martha had become quite sexless. Perhaps because of this, his depression actually worsened in the years after they married. To cope he was forced back to the survival strategies of his childhood — self-control, intellectual detachment and inner containment — elements that comprise the spine of psychoanalytic thought. But there was more to this, because it seems that the emotional distance that crept up between them was largely Freud's doing. In a telling comment about this Martha said, 'If I did not realize how seriously my husband takes his treatments I should think that psychoanalysis is a form of pornography' (Breger, 2000, p. 92). Biographer Ronald Clark has written that, 'during the 1930s there was even a slight irritation — well-controlled as Freud's reactions always were — and there was little left of the earlier great love' (Clark, 1980, p. 305). It seems that Freud was unable to sustain an equal, intimate relationship and in time he withdrew more and more into his world of books and ideas.

Freud had sincerely loved Martha, writing her long love letters during their protracted courtship. She had qualities that he found especially attractive, including a willingness to commit herself to domestic life without placing difficult demands on him. She was small, shy and obedient, as safe as a woman could be, but he seems to have erased her individuality and turned her into a kind of friendly servant. The marriage was never going to be an equal partnership, and with Freud increasingly throwing himself into his work, which it seems Martha never cared for, the inequality between them grew. This meant that she was unable to become the kind of intimate comrade who might have challenged and corrected him while also supporting him. Biographer Louis Breger comments that 'unrecognised conflicts caused a personal tragedy: he "lost the illusion" of a sexually pleasurable marriage and did not gain a vital and intimate partnership' (Breger, 2000, p. 95). Such a relationship was known to him only intellectually; he could imagine it and value it, and even remark that ultimately it was love that cured, but it was not the kind of marriage his parents had modelled and he had few tools for attaining it.

Perhaps another indicator of Freud's disillusionment was his relationship with his daughter Anna. She seems to have felt his distance within the family while she was growing up more acutely than his other children, and so she set out to have in adulthood what she had missed in childhood. She devoted herself to him and his work, never married or had children of her own, and in time became the guardian of his legacy. Freud analysed her himself, something that today would be considered incestuous, unethical, counterproductive and probably damaging, and that even back then raised eyebrows. If he ever had problems with this he nevertheless felt that he was able to transcend them. It beggars belief to imagine that he was unaware of the more important issues in such an arrangement, but neither he nor she ever expressed any reservations about it. Effectively, he allowed himself and his work to deprive her of any independent emotional life, and he did so knowingly. It is most likely that a kind of compassionate resignation was driving his side of this relationship.

Freud eventually became somewhat disillusioned with psychoanalysis itself (his published case studies are mostly records of failure). Quite early in his career he had cautioned that perhaps the most that could be achieved with some patients was to transform 'hysterical misery into common unhappiness', but in one of his last essays, titled 'Analysis Terminable and Interminable', he advanced several reasons why analysis sometimes did not work (Freud, 1937). These reasons included that perhaps the quanta of neurosis was just too strong in some cases, or maybe that the power of the death instinct undermined therapy, and he proposed several idiosyncratic personality types who were resistant to psychoanalysis. Some individuals had an 'adhesiveness of the libido', or they suffered a 'depletion of plasticity', or they were in the grip of 'a force which is defending itself by every possible means against recovery and which is absolutely resolved to hold on to illness and suffering' (in short, bloody-mindedness), to explain analytic failure.

This suggests that in some corner of his mind Freud must have guessed that psychoanalysis did not really work very well; (Theodore Millon advises that 'many narcissists have a degree of insight into their situation' [Millon et al., 2004, p. 353], and Freud was certainly extremely insightful, intelligent and reflective). When psychoanalysis appeared to succeed it was often because of quite other reasons than the ones that he had proposed. Certainly he had his successes; he cured composer Gustav Mahler of impotence during a lake walk in which they discussed Mahler's problems, and after a few instructions he cured conductor Bruno Walter of 'professional cramp' in his arm. But his successes did not conform to his theory of psychoanalytic cure, and his successful treatment of Walter showed him resorting to suggestion and charismatic influence and 'playing the magical healer' (Breger, 2000, p. 185) in a way that was quite contrary to his instructions for practice.

There exists one very good, independent account of a successful treatment by Freud, and it was written by a former patient, Albert Hirst. In this case a much more conventional picture of Freud the counsellor emerges. Hirst was a young man seeking a wise elder and he idolised Freud, who in turn liked him and was able to empathise with his sexual problems. Freud gave wise advice and was decent and caring, praising Hirst's intelligence and strength and urging him on in his efforts. This was orthodox humanistic therapy based on empathy, positive regard and congruence, and its key tool was the quality of the relationship that the two men formed. In this treatment Freud used all the key ingredients that research has shown to be most effective in therapy, but this was utterly at odds with his technical recommendations for how analysis ought to be conducted (Breger, 2000; Lynn, 1997).

Freud came to a theory of the relationship as the backbone of effective therapy when he formulated his notion of 'transference', in which patients transfer onto their therapists emotions and ideas and behaviours that derive from experiences with previous significant figures in their lives, typically parents. By 'analysing the transference' that developed in the therapeutic encounter, he hoped to be able to reveal the original dynamics that underlay the patient's disturbance, thereby using insight to cure. He arrived at this position in 1912, and subsequently revised it several times. It was a profound attempt to reduce cure to a single explanatory mechanism and, by and large, empirical research has borne him out in that the quality of the transference (the 'therapeutic relationship' or the 'alliance' in other terminologies) has indeed been shown to be the most important factor in successful therapy (aside from the basic commitment to therapy by the client).

But Freud used another tool to help at least some of his patients. Many people who associated with him have remarked upon his intense presence, his penetrating gaze and powerful aura. In fact, Freud cured by charisma; his mesmeric eyes and sympathetic voice, his empathic demeanour and inspirational vision, his suggestive words and his rock-solid faith in himself and his theories, combined to generate the levels of hope and confidence in his patients that enabled them to solve their problems, to attempt things that they would not otherwise attempt, to dream the impossible and to raise themselves up.

Clearly Freud's values and ideals had become much more pessimistic. He came to have an especially low opinion of humanity, at one point writing to a friend that most people were trash (Jones, 1955). Ferenczi summarised the change in Freud when he wrote in his diary:

> I think that in the beginning Freud really believed in analysis, he followed (Josef) Breuer enthusiastically, involved himself passionately and selflessly in the therapy of neurotics, lying on the floor for hours if necessary next to a patient in the throes of hysterical crisis. However certain experiences must have first alarmed him then left him disillusioned ... [Now] Freud no longer likes sick people ... He said that his patients were only riff-raff. The only thing patients were good for was to help the analyst

make a living and to provide material for theory. It is clear we cannot help them. This is therapeutic nihilism ... He loves nobody, only himself and his work ... He no longer loves his patients. He has returned to the love of his well-ordered and cultivated superego ... This involves, obviously, a degree of fatalism ... (Ferenczi, quoted in Masson, 1990, pp. 115–133)

Towards the end of his life Freud became increasingly preoccupied with promoting psychoanalysis. He was forced to flee Nazi Germany, and he settled in England in 1939. He died there that year in an assisted euthanasia. His four sisters were murdered by the Nazis in the 1940s.

In 2005 the book *Martha Freud: A Biography*, by Katja Behling appeared. Although like most biographies of Freud written by analysts it presents an uncritically sympathetic viewpoint, intriguingly it also shows that both sides of Freud's family — his own and his wife's — had financial failure and even criminality in their recent past. Martha's father became insolvent and spent time in prison, something she never spoke about, and Sigmund's father was rumoured to also have skated close to the legal edge in his own business failure. At another time Freud referred to his half-brothers' business dealings as involving 'forged money and credit papers' (Behling, 2005, p. 14).

The book also sheds significant light on Freud's out-of-role behaviour. It seems that when he was not playing the great leader he was basically a gentle if remote fellow. He told Martha that what he was most thankful to her for was that their children turned out so well (his son Martin wrote in his memoirs that he and his brothers and sisters had enjoyed very happy childhoods). He wrote movingly about Martha giving birth, saying that he had always 'treasured the priceless possession he had acquired in her', and that he had never 'seen her so magnificent in her simplicity and goodness as on this critical occasion, which after all doesn't permit any pretences' (Behling, 2005, p. 110).

Sigmund and Martha Freud's relationship illustrates typical themes in the charismatic marriage. Although Martha lacked confidence when young and never pushed the boundaries of her relationship with Freud, she was no mere doormat or weak complainer, her daughter Anna saying many years later, 'My mother observed no rules, she made her own rules' (Behling, 2005, p. 27). Certainly Sigmund dominated, but there was sufficient room within the marriage for her to grow and, if not to challenge him, nevertheless to achieve her own autonomy and integrity within her own values. The reason why they endured such a protracted engagement was because of the dominance of Martha by her mother, something that Sigmund was at pains to combat and then to replace with his own domination, but not so thoroughly as to completely stifle his wife. She seems to have been less his soul mate and more an honoured assistant. Behling sums their marriage up as:

> Freud demanded conformity and loyalty, and from the outset sought to mould Martha according to his wishes … It seems he saw in his wife an almost celestial being who protected him from any kind of meanness, and it took him greatly by surprise when she once admitted that at times she had to suppress bad or evil thoughts … His mild forbearing manner was that of a man who did not hold her ignorance and simple-mindedness against her — for she was fortunately so reasonable in other respects. It seems he took it for granted that he would emerge victorious from whatever game they played. In this way Martha would be a constant reminder of his greatness and superiority. At the same time however she was not to be too meek. He wanted to goad her into opposition, to provoke her in order to then lavish her with affection. (Behling, 2005, pp. 78–80)

Perhaps the strangest question remaining about Freud concerns his astonishing lack of insight into himself. He was, after all, the greatest psychologist of history, so he might be expected to turn his questioning on himself and he did this in his self-analysis. He came to doubt the efficacy of psychoanalysis, but he never questioned the personal myth he created in his self-analysis; he never came to see himself as a charismatic prophet intent upon remaking the world by sheer creative willpower, with all the strengths and weaknesses that went with that role. This is the norm for charismatic personalities; they can be uncannily perceptive of others but quite blind to their own motives, yet sometimes they reveal just a hint of insight. Just as great system-builders seldom include themselves in their models, usually placing themselves as great exceptions outside their own theories, so psycho-analysis was driven by sexuality, but Freud preferred to see himself as a man of reason and self-control. Perhaps some truths are just too painful to face.

Endnotes

1 See especially: Breger (2000), Ellenberger (1970), MacMillan (1991); Sulloway (1979), and Webster (1995).

2 Theodore Millon advises that, 'parental overvaluation through the stages of neuropsychological development may be seen as the core factor in the development of narcissistic patterns' (Millon et al., 2004, 358).

3 See Freud's 1916 essay, 'Some Character Types Met With in Psychoanalytic Work'.

CHAPTER FIVE

· · · · ·

. . . . .

Perspectives From
Psychology and Psychiatry

*When a superior intellect and a psychopathic temperament coalesce —
as in the endless permutations and combinations of human faculty
they are bound to coalesce often enough — in the same individual, we
have the best possible condition for the kind of effective genius that gets
into the biographical dictionaries. Such men do not remain mere
critics and understanders with their intellect. Their ideas possess them,
they inflict them, for better or for worse, upon their companions or
their age.*

William James

The above quote from William James, the father of American psychology, suggests an explanation of charisma. If mental illness is combined with genius it may lead to greatness. But what kind of pathology, and what kind of genius for that matter, did James have in mind? Full-blown insanity would seem to be incompatible with sustained creativity or leadership.

It turns out that almost every type of psychopathology has been associated with genius of some kind. The Nobel Prize-winning mathematician John Nash, subject of the film *A Beautiful Mind*, had schizophrenia. Composer Robert Schumann suffered from bipolar disorder (or manic depression). Depression afflicted several political leaders of genius, notably Abraham Lincoln, while sociopathy seems to have driven Joseph Stalin. Schizotypal and schizoid disorders have challenged several artistic geniuses including Salvador Dali and Andy

Warhol. Both Albert Einstein and Isaac Newton probably suffered from Asperger's syndrome. At a slightly less-elevated level, Howard Hughes suffered from obsessive–compulsive disorder, and many high achievers have suffered from attention deficit and/or hyperactivity disorder. The book *Divine Madness* by Jeffrey Kottler (2006) presents profiles of several highly creative people who were at times quite mad. But when considering only those individuals who are described as charismatic, the category of 'borderline personality', and especially that subtype known as narcissistic personality disorder (NPD), is the pathology most frequently assigned to them.

Psychiatric tradition recognises four levels of mental health on a continuum from healthy to insane. The healthiest is 'normal' and indicates someone who is realistic and functional all of the time and who shows little or no psychological disturbance. As a technical term (and for obvious reasons) this label is seldom used today, although it has by no means been entirely abandoned.[1]

The next point on the continuum is 'neurotic', which implies a modest degree of disturbance such as describes most of us most of the time and some of us all of the time. Such disturbance may include frequent or occasional irrationality, behavioural dysfunction, anxiety or depression, compulsions and obsessions, substance abuse, or any of the myriad other indicators of mental illness, although the sufferer may nevertheless hold down a job, maintain a marriage and raise a family without significant disruption.

Then comes the 'borderline' position, which describes persons on the border between sanity and madness. Such individuals appear to be merely neurotic most of the time, but frequently they exhibit behaviour that is extremely odd. The borderline personality may not actually hear voices or suffer delusions (though some do at times), but their behaviour often deviates greatly from social norms.

Finally at the most disturbed end of the continuum is the psychotic, typically the schizophrenic or manic depressive.

The borderline position on the psychiatric continuum includes what are called 'personality disorders'. The *Diagnostic and Statistical Manual of Mental Disorders, Fourth Edition, (DSM IV)*, the textbook of American psychiatry, describes 12 such disorders; other systems recognise a few more. A familiar example is the antisocial personality disorder or psychopath, the person who is hostile, deceitful, aggressive and lacking in conscience; at worst, the serial killer who may also be charming and functional in most other ways. Another example is the paranoid personality disorder, the person who is highly suspicious, anxious, untrusting of friends and associates, and who reads all sorts of hidden meanings and threats into the innocent remarks of others; typically the conspiracy theorist. A third example is the dependent personality disorder, the person who has difficulty making decisions on their own, who needs others to lead them and who may be led into dreadful behaviours by unscrupulous leaders, as with the concentration camp guards who insisted that they were 'merely following orders' when gassing prisoners. All these disorders may present with a background of apparent normality.

Individuals suffering from personality disorders are not insane in the usual sense, but they are unstable; under severe stress they may display psychotic symptoms. They show a pervasive and enduring pattern of consistently disturbed thought and behaviour that differs markedly from normal behaviour, and which is much more complex and severe than mere eccentricity. It is crucial to appreciate that while we may all behave like some of this on occasions, the individual with a personality disorder is a 'borderline' and thus is unstable and veering towards madness. Thus he or she behaves like this so frequently, so unrealistically, so extremely and so irrationally, as to cause others to suspect that there is something distinctly odd about him or her.

In clinical settings, personality disordered patients are striking. They may confuse even experienced practitioners with their manipulations. They are often highly creative and deceptive; probably they have had to be to survive. They show extraordinary 'fragmentation' where one or more parts of their personality are 'split off' and behave in perverse and contradictory ways. Yet they may also appear quite 'normal' — warm, friendly, persuasive, realistic and reasonable. Consider the following vignette.

> Paul came to see me in my fourth year of practice. He was a professional man in his late thirties, well-dressed, well-spoken, fit and tan. We chatted amiably and he seemed

confident and remarkably buoyant I thought, given that he had presumably come about some personal problem. Time passed, and after a very pleasant twenty minutes I still had no idea why he had come, so I asked him directly. He said that he wanted a psychological assessment; he had been unfairly dismissed from his government job and he was suing his former employer and he needed a psychological report saying he was competent to return to work. I accepted this and agreed to help him. I gave him a questionnaire to take home to complete, and booked him in again next week. Then, suddenly aware of the time, I rushed through the last twenty minutes by taking his history. He seemed to have had an uneventful life.

Next week he was again warm and casual, and again we chatted freely. I took his test answer sheet and continued with the life history, getting the names and contact details of persons who might verify his statements. He provided all this to me along with details of his unfair dismissal. I concluded that he had been treated abominably by a petty tyrant in the public service. Only at the very end of the session did he provide a detail that I found odd; he had actually been dismissed from his job five years ago. I had assumed that his dismissal was recent. Now I was so flabbergasted (and time was up) that I didn't press him on this.

After Paul left, questions flooded my mind. Why hadn't I questioned him closely from the start? Why had I succumbed to his time-wasting? Was it because Paul was so good at disarming people, or was I just incompetent? Had he deliberately delayed discussing his background until time began to run out? Why? How had he created in my mind the notion that I was dealing with some current issue when he'd lost his job five years ago? Maybe I hadn't really listened? The more I thought about Paul the more confused I became, and the more I doubted myself.

I sent the test off and phoned the names he'd given me to verify his account of things. The first person was a friend who verified everything Paul had said. Great, I thought. The next was a woman whom I'd assumed was another friend, but she turned out to be Paul's fiancée. He'd not mentioned being engaged, but at this time I doubted myself so severely that I assumed that he had and I'd forgotten. She told me a very different story, about Paul's stubbornness and the difficulties she felt in their relationship. They'd been engaged for almost six years and he'd put off the wedding many times; she'd actually given him an ultimatum, but in the last two years their sex life had stopped and now she wasn't sure she wanted to get married at all. Bitterly, she told me that Paul had become so obsessed with his legal case that he'd lost most of his friends, hadn't worked for two years, drank heavily (he'd denied this to me), and even his health was suffering. Astonishingly, I'd picked up none of this in my discussions with him.

I got the test results back from the testing agency that afternoon. Paul's responses were invalid because he had 'faked good'; he had falsified his answers to present a good image of himself, and the test had identified this.

Next session Paul happily breezed in again. Now I ignored his friendly chatter. He was nonplussed when I told him of the test results, and he happily offered to take the test again; problem solved, he suggested. He smiled broadly when I reported what his fiancé had said; he didn't take her protestations seriously. He had only wanted her to confirm the details of his job loss; anything else was irrelevant, he said. Now I felt very uneasy, so I said, 'Paul, I don't think I can continue with this. I don't think you're being honest with me. I don't think you were honest doing the test, and I think you've got much bigger problems than the legal matter, your relationship for example'.

Paul's face clouded, then he exploded: 'You bastards are all the same. That's what the last two shrinks told me. Why can't you get it through your fucking heads that all I want is an assessment that shows I can work. I've told you that, several times. You don't listen, you get confused, you jump to conclusions, you don't understand anything I say, you waste my time with your psychological bullshit, then you bill me for it and you won't do what I ask'. Waving his fist, at me he stormed out.

Several things stood out about Paul's behaviour. First, his easy amiability was very seductive but it concealed a deep hostility. Usually I can tell angry clients from their tight body language, but not Paul. He was jovial and seemed keenly interested in me, the world and others. But he must have smouldered underneath.

Second were his contradictions: his amiability and his hate, his engagement to his fiancée and his disregard for her, his deception and his self-righteousness, his pursuit of justice and his amorality. He obviously saw nothing amiss with withholding the fact that he'd previously visited two other shrinks, but he held me to a different standard. These contradictions seemed to co-exist comfortably within him without modifying each other, as if they were split-off and not integrated into his psyche.

Third was his influence on me. At our first meeting I'd been so mesmerised by his warm banter that I'd neglected my professional duty. Later, I became so confused by him that I was almost unable to function. Finally, I'd felt utterly intimidated by him, shocked and anxious. Such powerful influence, for good and ill, just seemed to emanate from him. I doubt if he knew the full effect he had on others; he just did his thing, but he could be devastating.

Finally, when confronted, Paul seemed to have at the ready a brilliant defensive armoury. He had precisely identified my self-doubts and turned them against me, accusing me of not listening, of being confused, of not understanding, and of jumping to conclusions, all of which were at least partly true. Then he'd blamed me for the time-wasting, something I already felt guilty about, and implied that I was a money-grubbing fraud. He unerringly went straight to my most sensitive spots. He, of course, was righteously indignant.

Paul probably suffered from narcissistic personality disorder. Some of his behaviours, his defensive strategies for example, were brilliant and deliberate, but other of his behaviours such as his contradictoriness, his certitude and his stance of total moral superiority, exposed his disturbance. As William James proposed, sometimes genius can co-exist with madness. Paul was quite dysfunctional, but others like him may be quite functional; intensely observant, highly creative, brilliantly entertaining, seductively warm, powerfully manipulative and covertly controlling.

Narcissistic personality disorder is one of the less common of the borderline conditions, afflicting less than 1% of the population (Semple, Smyth, Burns, Darjee, & McIntosh, 2005, p. 453). It is defined in the DSM IV as involving at least five of the following behaviours:

1 *Grandiosity — an exaggerated sense of self-importance.* The narcissist 'knows' they are superior to others and expects to be recognised as such even without proof.

2 *Preoccupation with fantasies of success, power, brilliance, beauty or ideal-love.* The narcissist may live in an inner world in which they already have the recognition they believe they are entitled to.

3 *A belief that one is special, unique, or a member of an elite.* Such a person may believe that they ought to only associate with others on their elevated level, and can only be understood by them.

4 *Needing excessive admiration.* Typically such individuals recruit others as uncritical followers to bolster their self-esteem.

5 *Having an inflated sense of entitlement to the favours or compliance of others.* This arises from their grandiosity and the expectation that they will be recognised as superior and special.

6 *Exploitative and manipulative.* The narcissist uses others to achieve desirable ends for him or herself, typically by concealing their narcissism and posturing as a leader and controlling others covertly.

7 *Lacking in empathy.* The narcissist does not recognise others as equal to himself, hence he cannot place himself in their position or identify with their needs and feelings.

8 *Envious of others, or believes that they are envious of him/her.*

9 *Arrogant, haughty demeanour or attitude.*

In order to be classified as suffering from narcissistic personality disorder, an individual must exhibit 'clinically significant distress or impairment in social, occupational, or other important areas of functioning' (APA, 1994, p. 275). Hence, the diagnosis only applies to those persons so disabled by narcissism that they can barely function. But there also exists a functional variant. It occurs when high intelligence and effective social skills, self-discipline and talent enable the narcissist to function well in the world, to adapt and cope, to conceal the extremes of their behaviour, and perhaps even to develop them in extraordinary ways (Masterson, 1988).

Such a person is forced to become a sophisticated student of human behaviour in order to survive with their narcissistic mindset intact, and to turn their narcissistic traits to their advantage. A preoccupation with success can lead to continual strategising about how to achieve it, and a commitment to the necessary hard work and to developing talents for leadership and rhetoric. Aloofness can be detachment, leading to the 'helicopter effect' — an elevated overview of the social world, a sharp perception, and cool level-headedness. A sense of oneself as the centre of the universe may lead to fearlessness, which in turn may provide acute insights into the fears of others. Such fearless detachment may enable the high-level narcissist to accurately diagnose problems and to strategise solutions, and to confront others and demand their compliance in ways that more realistic personalities would not dare. A preference for remembering the past as one would have liked it to have been rather than as it was (Millon et al., 2004) may give one freedom from emotional 'baggage', and this can seem liberating to others (though the narcissist has his baggage all right). A need for admiration may force him to develop the social skills required to get it. Grandiosity, in the form of a certain jaunty puckishness, can be very winsome. Envy, resentment and lack of empathy may lead him to critically engage with those who do not automatically recognise his 'superiority'. From this may come a strategic scrutiny of them, which may in turn lead to an accurate diagnosis of their flaws and weaknesses, the better to recruit them to subordinate dependency. Such individuals are able to enlist others to act out their fantasies, and to get the recognition and admiration they feel they deserve.[2]

They may also be great in a crisis; their detachment and fearlessness can be crucial strategic tools. Consider a highway smash. Most witnesses of crash scenes are horrified, sensing their own fears and vulnerabilities in the hurts of those involved. They want to avoid, to drive on and to remain uninvolved, as if to deny their own mortality. They risk feeling overwhelmed by their emotions, and have to force themselves to stop and provide assistance. Or, some may become ghoulishly fascinated, as if on the edge of an abyss, mesmerised by the dreadful mortal truth that has opened up before them.

The narcissist is likely to react differently. He is able to help the injured and comfort the grieving with appropriate care and sensitivity because his emotions do not intrude. He can be calm and competent because he does not really see himself as like these others. For all his apparent (even genuine) compassion, his stance is more like that of a kindly vet treating an injured animal, than of one human being helping a fellow sufferer. His detachment enables him to respond effectively and efficiently to the victims without experiencing the horror and revulsion that others feel, or at least not to the same degree. Obviously, as an intelligent person he is consciously aware that such a tragedy might happen to anyone, but he is sustained by an unconscious grandiosity that holds him aloof. Intellectually he knows it could happen to him, but emotionally he knows it will not!

Functional narcissism is quite different to the self-serving illusions we all have, those secret beliefs we harbour that we are in some small way better people than those around us, more deserving or decent, and that our families and children are special and somehow more lovable than other families and children. Such beliefs are quite normal, the remnants of primary narcissism in us all, defensive projections of self-esteem. It is only when we start acting on them in a consistent and aggressive manner that we begin to approach pathological narcissism.

Narcissism involves something much deeper and more wide-ranging than mere egoism, or even extreme egoism. It is also much more than just selfishness or self-absorption or self-centredness. It involves a profound shift in one's perception of the world, and also of one's relationship to it. A good illustration of this comes from an anecdote told by psychiatrist James Masterson about a narcissistic client he was treating. As Masterson tells it, he was intending to go on holiday and he gave his clients several weeks' advance notice (some patients find it distressing to have to suddenly break of treatment then resume it again on someone else's timetable). But one particular man made no response when initially informed, nor for several sessions after. Only towards the end of the last session before Masterson was about to leave did this chap tell him: 'You know, I've been thinking about this holiday you're going on, and I'm a bit concerned. You see, these sessions of ours are the high point of the week, and frankly, I'm not sure how you'll manage without them'.

The confusion of self and other revealed in this statement is not only typical of narcissists, it also illustrates the fundamental difference between narcissism, egoism and sociopathy — three conditions that have much overlap, yet which are distinct. An egotist knows that other persons exist; he just thinks that he is better than they are. A sociopath also knows that others exist, he just does not care about them. A narcissist, however, in some strange way does not fully grasp the reality of the existence of others; perhaps intellectually he does, as ideas, but without emotional weight and colour, and without empathy. They exist as shadowy images to him, as reflections of his own thoughts and feelings.[3] Yet despite this delusional quality, narcissism is not merely a delusion. A delusion is an irrational idea that takes hold of an otherwise normal and rational mind because it serves some neurotic or psychotic purpose. Narcissism is really about not having certain key ideas in the first place. It is an absence rather than a perverse presence.

Another illustrative anecdote comes from the book *I Know You Really Love Me*, by psychiatrist Doreen Orion (1977). She was stalked by an obsessive patient who fell in love with her (erotomania), and she wrote about her experience and researched other cases. In one story, an obsessive male stalker was finally confronted in court by the district attorney, the police, his victim, lawyers and the judge, and forced to accept that his victim really did not want to receive his telephone calls anymore. Rather than collapsing into despair, he burst out with, 'Well, fuck her then, she can call me'.

Such accounts show an inability to feel weakness, or distress or vulnerability of any kind, and the tactic of projecting such feelings onto others. The psychiatric patient is 'concerned' for the psychiatrist. The stalker angrily demands that from now on his victim must do the phoning in their 'relationship'. There is poor awareness of boundaries, a complete lack of recognition of their motives, and an inability or refusal to consider these. It is not so much that there are bad intentions, although these may also be present; rather, it is that there is a piece missing in the narcissist's mental apparatus, a fundamental incomprehension of the true otherness of others. Because the narcissist sees the world and others as an extension of their own self, there is an identification of the needs and feelings and vital interests of others with their own.

Other traits that successful narcissists possess deserve mention. First is enormous energy; some work long hours, seven days a week, in order to overcome with sheer energised determination the obstacles that they encounter. They have such energy because of a lack of awareness of inner conflicts. The doubts and uncertainties that plague ordinary people are repressed by narcissists in their phobic denial of their vulnerabilities. There is no draining inner tension, and others find such clarity refreshing and attractive; one writer referred to this as the 'fascination of the unconflicted personality' (Olden, 1941, p. 348). Of course there are inner conflicts, quite severe ones, though well-hidden, but staying hyperactive is a way of avoiding such conflicts, outrunning them and thus denying their existence. Without such an escape the person might lapse into depression.

They are also unable to attach themselves to others with any depth or genuine connection. Their relationships are shallow because they are unable to internalise others or to invest themselves in them. They see others as means to their ends, rather as persons in their own right, and thus are unable to participate in equal friendships. Because of this their relationships are either short-lived or, where they endure, this is likely to be because their partner is very dependent, or is willing to tolerate an unequal relationship.

But most of all, such persons are unable to just be, to commune as an equal, to be thought of as ordinary and to acknowledge vulnerabilities. They are constantly attuned to issues of power and control, and are driven by a need to be the centre of attention, to be 'on-stage' the entire time. They may suppress this need with self-discipline for a while, or they may relax on holiday and fall asleep in the sun for an hour or two, but it always re-emerges.

Of course, we can all behave like some of this on occasions. But for most of us, when we do so we feel awkward, and afterwards can be forced to admit that we went too far, were out of line or over the top. What makes the narcissist different is that they have no such conscience or awareness. They sincerely believe that their behaviour is realistic, rational, moderate and even for our own good (if only we would realise it), and that any admiration or compliance they get from others is only their due. They behave like this so consistently that they are deemed to be 'on the borderline' of madness.

This is the paradox of the charismatic personality. Such individuals are clearly narcissistic and severely so, yet because of their brilliance they are able to function, and even to excel. They are not distressed or impaired; they work and marry and succeed, and they may even have a touch of genius that enables them to achieve great things. Hence, to describe a charismatic leader as narcissistic is appropriate, but it is a mistake to imply that this is all he or she is, for the reality goes much further.

This situation is further complicated by the fact that personality disorders are not discrete; secondary disorders exist and are highly relevant. Alongside their narcissism the charismatic personality may show a range of other borderline traits such as antisocial behaviour, passive-aggression and paranoia.[4] Some charismatic religious leaders appear to live in something of a fairyland, suggesting schizoid tendencies. Even depressive (Stevens, 1998, p. 42; Stevens & Price, 1996) and dependent tendencies may keep the charismatic personality aligned to others whom they need for narcissistic supply. As Charles Manson once told an interviewer, 'Something in me needed them much more than they needed me' (Emmons, 1988, p. 183).

Lastly, there is a distinct possibility that psychological disturbance may not detract from creativity but may actually enhance it. Perhaps the 'psychopathic temperament' that William James refers to really does enable the charismatic personality to see further than others, or more clearly into the heart of things. Perhaps visionaries and prophets really do see something that the rest of us are unable to, if only we (and probably also, they) could understand it. As Edgar Allen Poe put it:

> Men have called me mad, but the question is not yet
> settled whether madness is, or is not, the loftiest intelli-
> gence, whether much that is glorious, whether all that is
> profound, does not spring from disease of thought, from
> moods of mind exalted at the expense of general intellect.
> (cited in Ostrom, 1948)

The psychiatrist who has had the most to say about narcissistic disorders is James Masterson, author of several books on the subject
including *The Search for the Real Self* (1988). Masterson supplements
psychoanalytic notions with the concept of the False Self, a persona
that he says narcissistic individuals construct to ease their way in the
world. This idea originated with William James, and is analogous to
the existential notion of 'authenticity' (May, 1973; Rogers, 1995). It
was also taken up by the humanistic psychologists Abraham Maslow
and Fritz Perls, who espoused the importance of 'being real' to
oneself (Maslow, 1968; Perls, Hefferline, & Goodman, 1951).
According to Masterson, the Real Self is

> Made up of the sum of the intrapsychic images of the self
> and of significant others, as well as the feelings associated
> with those images, along with the capacities for action in
> the environment guided by those images. The images of
> the real self are derived mostly from reality ... and its
> motives are directed toward mastery of reality ...

In contrast the False Self is:

> Derived mostly from infantile fantasies, and its motives are
> not to deal with reality tasks but to implement defensive
> fantasies ... the purpose of the false self is not adaptive but
> defensive; it protects against painful feelings. In other
> words, the false self does not set out to master reality but
> to avoid painful feelings, a goal it achieves at the cost of
> mastering reality. (Masterson, 1988, p. 23)

There is a splitting of the person implied here. Under pressure from
the environment the child divides its experience into two parts: a
troubled but realistic part from which the only conclusion to be
drawn is that the world is a painful and dangerous place; and a
fantasy part in which the evil of the world is subsumed by a saving

fantasy that elevates the child and transcends the pain it experiences. The realistic part is then spilt off and denied, insofar as this is possible, while the fantasy part — the false self — is nurtured and elaborated. Masterson adds:

> It is in the nature of the false self to save us from knowing the truth about our real selves, from penetrating the deeper causes of our unhappiness, from seeing ourselves as we really are — vulnerable, afraid, terrified, and unable to let our real selves emerge. (Masterson, 1988, p. 63)

According to Masterson, the false self enables the narcissist to manipulate others to perform important roles for him; to feed his grandiose self-concept by gaining their applause, alliance or even worship. As long as these others feed the false self with such narcissistic supply he feels inflated and secure. If the necessary applause is not forthcoming the narcissist may become depressed or, more likely, plunge into a frenzy of activity aimed at restoring this supply.

Narcissistic supply comes through mirroring in which the narcissist looks at others in his environment, and to the environment itself — his family, possessions and other symbols of virtue, success and excellence — to reflect his sense of grandiosity and specialness. But it takes much hard work to validate the narcissistic claims of the false self, and even the attainment of these claims fails to satisfy in the longer term. So new dreams must be erected, perhaps eventually leading to the proclamation of a great problem-solving program such as a messianic or revolutionary movement. Ultimately, only being God can really satisfy the narcissist, and because this is impossible he remains eternally frustrated.

The mirroring that the narcissistic personality requires can be subtle and mysterious. Often it is not merely blatant hero-worship; it can disguise itself in all manner of devious ways. Masterson tells an anecdote which illustrates this:

> Several years ago ... I was quoted in a *New York Times* article about the major symptoms of the narcissistic personality disorder. Within days I received twelve phone calls from individuals who had read the article and suspected that they might have the disorder. Each came to see me for an evaluation, which as it turned out proved them right. They were

> narcissistic disorders and on finding out they each asked if they could come for treatment. I agreed, but since I didn't have time in my schedule to take on twelve new patients, I suggested they see one of my associates in the Masterson Group. Not one returned to begin treatment, which indicated they were not truly interested in treatment for its own sake, but rather were looking toward me as a narcissistic supply. Because of my reputation in the field and because I was quoted in the *New York Times*, being in therapy with me would reinforce their grandiose images of themselves, whereas being in therapy with anyone else would be seen as a weakness. (Masterson, 1988, p. 95)

Another function of the false self is denial; the narcissist denies vulnerability, need, error, and suffering, and attributes his problems to the failings of others. If his wife leaves him he gets angry rather than sad; he is right, her quibbles are trivial and she is ungrateful. In his dealings with others he makes plain that he is guiding and directing them for their own good. If possible he works in a profession such as medicine, counselling, politics, acting or modelling where he can get narcissistic supply with little effort. If he achieves this he may live for years denying the emptiness of his life.

Perhaps because of his scientific training, Masterson is less inclined than the psychoanalysts to speculate about the genesis of the narcissistic personality. He accepts that it probably involves a developmental arrest very early in life, between about 18 and 36 months, but adds that how and why this arrest occurs is not always clear (Masterson, 1988, p. 101).[5] Masterson also describes the important category of what he calls the 'closet narcissist', who does not feel bold enough to express grandiosity openly (though they feel it), and so identifies with another person or group through which to indulge their narcissism. The importance of this category is to demonstrate that high-functioning narcissists are invariably bright and adaptable, and that they can adjust their behaviours to fit in with whatever social realities they face. The flagrantly exhibitionistic egoist is only an extreme, and somewhat caricatured, variant of the narcissistic personality. More commonly, such a person seems at first to be merely charming, colourful, smart and passionate, and only gradually does the full extent of their borderline condition unfold.

But the false self is not entirely false, for it solves the problems of the time when perhaps nothing else can. It is in this way legitimate, and it may become more so with time. As anyone who has had to learn a second language can attest, it is possible to take on a body of knowledge with little initial appeal, or even a role, and through application over time, master it and come to enjoy it so that it feels quite natural. Indeed, anything that one throws one's heart and soul into may, at some level, come to represent or symbolise something deep in one's nature. It may even be that in order to be viable, the false self must express something of one's deepest being (what Heinz Kohut calls the 'agenda of the nuclear self', and which will be discussed further in chapter 8). Perhaps after all, the false self is quite 'normal,' as Houselander has argued:

> Man cannot face himself as he is. In spite of the fact that this is an age in which introspection has become almost a science, it is also an age in which few men dare face themselves and to see themselves as they really are. Consequently nearly everyone has made a fictitious self, and in contemplating this, he is able to forget 'what manner of man he is'. (Houselander, 1952, p. 49)

This raises the issue of whether the false self is really all that different from the flattering myths we all create about ourselves in daily lives, and which constitute a supportive narrative and comforting self-image for each person. As Ernest Becker has shown, the great aim of each individual is to cast him- or herself as a hero in their own life (Becker, 1962). Perhaps the false selves of narcissistic personalities and charismatic leaders are really just quite normal defensive strategies writ large, although such persons do seem to cling more tightly to these personae than lesser people cling to their self-images.

Finally, the false self is not usually just a blindly heroic 'front'. Churchill's clearly was, but the false selves of other great leaders and achievers are not interchangeable. For example, Eleanor Roosevelt developed a false self that expressed her value as a helper; making herself useful in a supportive role enabled her to escape the pain of her upbringing (see chapter 12). The false self is developed in the context of the opportunities and information available to the person at the time of its construction, and may also reflect the individual's deepest sense of who they are; hence there are bound to be great variations between the false selves of those who present themselves to us as leaders.

It is clear from the cases of Churchill and Freud that something of the sort occurred in their development. Each devised a heroic fantasy of himself as special in some way, a false self that was superior to others and destined to achieve great things. Each was resourceful enough to make their fantasy work for them, transforming themselves into the image of the fantasy they had created, and living extraordinary lives along the way. Yet because of their adoption of false selves, sooner or later each faced problems reconnecting with others in a realistic, rather than a narcissistic way. This is the problem of intimacy.

Masterson argues that intimate relationships are the bête noire of the narcissistic personality (divorces are more common among narcissists; Millon et al., 2004). Intimacy may be doubly problematic for persons who have been able to amplify mere narcissism into a full-blown charismatic personality. For humans are social animals; we have a basic longing for connection with others no matter how alienated from our true selves we become, and no matter how disillusioned or embittered with humanity we may be. This is why, when surveys are done to study happiness in a population, repeatedly the happiest individuals turn out to be those in loving long-term relationships (Argyle, 2002). The rewards of such relationships surpass the baubles of narcissistic supply, and even exceed the satisfactions of health and longevity in their capacity to delight heart and soul. Although Armand Hammer famously said that his first and only love was money, most of us, charismatic personalities included, are not so constituted. This is the great challenge for charismatic personalities; as psychoanalyst Nancy McWilliams has advised, the most 'grievous cost' of narcissism is a 'stunted capacity to love' (McWilliams, 1994, p. 175).

Endnotes

1   *The Oxford Handbook of Psychiatry*, for example, has much to say about being 'normal'. See Semple et al., 2005.

2   A topic seldom mentioned in clinical writings is the impressive talents that borderline disturbance sometimes bestows. As Daniel Goleman, in his best-seller *Emotional Intelligence*, put it:

> While emotional neglect seems to dull empathy, there is a para-doxical result from intense sustained emotional abuse, including cruel, sadistic threats, humiliations, and plain meanness. Children who endure such abuse can become hyper-alert to the emotions of those around them, in what amounts to post-trau-matic vigilance to cues that have signalled threat. Such an obsessive preoccupation with the feelings of others is typical of emotionally abused children who in adulthood suffer the mercu-rial, intense, emotional ups and downs that are sometimes diagnosed as 'borderline personality disorder'. Many such people are gifted at sensing what others around them are feeling, and it is quite common for them to report having suffered emotional abuse in childhood. (Goleman, 1996, p. 102).

3   Carl Jung also has theorised about what he calls the 'Mana Personality' and his observations are worth considering although they will not be covered in depth here. The relevant passages in Jung's writings are: *Collected Works*, Volume 7, paragraphs 110 and 382; and Volume 12, paragraph 563. This overview derives from the work by Samuels, Shorter and Plaut (1986).

4   Through a confounding of nomenclature the term 'borderline' has been used in psychoanalysis as a name for the range of mental health and per-sonality functioning comprising those disorders existing between the neurotic and the psychotic ranges, and by psychiatry as the name for a specific disorder — borderline personality disorder — within that range. The situation is confusing and amateurish. See Millon et al., 2004, p. 480 for a discussion of this.

5   Coming from a more empirically-based research orientation, Bass and Riggio also agree that the roots of charisma probably begin early in life (Bass & Riggio, 2006, pp. 143, 145).

CHAPTER SIX

● ● ● ● ●

• • • • •

The False Self
of Adolf Hitler

Those to whom evil is done, do evil in return.

W. H. Auden. September 1, 1939

Several problems confront anyone attempting to understand Adolf Hitler. First is that he was constantly acting and his real self was probably never seen by anyone (Waite, 1977, pp. 452–454). Many commentators have failed to appreciate this and have been led to erroneous conclusions. But Hitler was also much more than just an actor; he was a highly complex character, brilliantly intelligent, charming and a polished liar. As well, he was a politician of conviction and fanaticism. The cultivated French ambassador Andre Francois-Poncet described him as follows: 'The same man, good-natured in appearance and sensitive to the beauties of nature, who, across the tea table expressed reasonable opinions on European policies, was capable of the wildest frenzies, the most savage exaltation and the most delirious ambition' (Waite, 1977, p. 37). Hence appreciating the psychology of Adolf Hitler presents a major challenge.

Second, he was profoundly disturbed. He was deathly afraid of water and he hated the moon. He had a fear of the dark, of being alone and he distrusted the passage of time; even on fine days he preferred to stay up all night with the lights blazing and the curtains drawn, entertaining associates to avoid being alone, only going to bed as the dawn approached. He was obsessed with the colour of his eyes, the width of his nose, the shape of his skull, the shape of his hands and the length of his fingers, comparing his hands to those of Frederick the Great. He was fascinated with fire, wanting to have an open fireplace in every room. He was compulsively clean, washing and scrubbing his face and hands after every meal, and shaving twice a day. He was both fascinated by, and terrified of, syphilis. He kept his body covered at all times, wearing long white underwear even on hot days.

Hitler was also a great hater; he had films made of the executions of criminals and political opponents, and he enjoyed watching them die. He was a prude but he had a taste for erotica. He was a hypochondriac. He preferred the cold; his tea parties were held in chilling temperatures. He had several superstitious beliefs that verged upon the delusional; for example, he believed that the magical elixir of youth really did exist and he contemplated sending an expedition to India to find it. He was prone to depression and very emotionally labile. He was highly self-pitying and had a strong streak of infantilism; he was addicted to sweets, taking seven teaspoons of sugar with his tea and claiming to consume two pounds of chocolates each day. He was also addicted to prescription medicines, and sometimes he sucked his little finger for comfort.

One of the most telling symptoms of his disturbance was that he was unable to admit a mistake. His press chief who followed him closely for 10 years recalled that if an associate corrected him on any point of fact, no matter how valid or minor, Hitler could not admit his mistake. His personal interpreter, who worked closely with him for 10 years, also said that he had never heard Hitler admit to an error, even to his closest friends (Waite, 1977). Hitler refused to speak to his private secretary for six months because she expressed doubt in his ability to know about the effect of alcohol on health, and he eventually claimed to be 'infallible' on political matters (Schweitzer, 1984). Indeed, so severely neurotic was Hitler that it has been argued that he consolidated his fears into a single object, the Jews, and this is probably so (Waite, 1977).

Third, hundreds of misleading books have been written about him, from apologia to crackpottery. The mythology surrounding this greatest-of-all mass killer is so extreme as to distort one's best efforts at penetrating him. For example, it has been proposed that he was a coarse, uneducated psychotic whose success in politics was a complete fluke. It has also been argued that he was a great leader who did not initiate the holocaust and may not even have known of its existence; had he died in 1939 he would be considered a great statesman. In sum, this is a polarised field, with only a handful of genuinely authoritative works on the personality of the man, and even these have major limitations.

Fritz Redlich's (1998) book *Hitler: Diagnosis of a Destructive Prophet* is perhaps the best, and is used here as the central guide. It is an attempt to present a thorough account of Hitler's physical and psychological health throughout his life. Written by an eminent psychiatrist with access to the extant medical documentation about Hitler, but ranging far beyond this, the study promises much. Unfortunately, psychiatry is itself a controversial field, with competing schools and internal divisions. At one pole are the psychoanalysts whose extreme subjectivism really does constitute a special lexicon, while at the other pole there are medico–biological determinists of various stripes for whom practically all human behaviour is reducible to biology. Occupying the middle ground are the proponents of the mainstream bio/psycho/social approach that postulates a complex interactive view of behaviour, but even within this there are many idiosyncratic viewpoints. Redlich's approach seems to reflect this; on two points crucial to the present perspective — whether Hitler may have suffered from a personality disorder, and his 1918 visionary experience — Redlich is dismissive rather than helpful, and he also gives short shrift to the diagnosis of psychopathy (Redlich, 1998, pp. 333–335). Indeed, so at odds with current research into psychopathology is Redlich's account as to render it quite misleading.[1] In addition, any book that purports to be a comprehensive medico–psychiatric study yet misses something as fundamental as Hitler's homosexuality, has to be considered incomplete.

Another central text is R.G.L. Waite's (1977) *The Psychopathic God: Adolf Hitler*, which attempts with great resource to cast him as a perverted psychopath. It too is filled with fascinating detail, but it also has numerous failings, as Redlich has pointed out. It is a systematic attempt to grapple with the psychology of Hitler and it has stood the test of time, as have several other books, including the volume by Haffner (1979) used herein; but too often Waite indulges in dubious psychoanalytic speculation.

As well as these two, credible new materials continue to emerge and these will be mentioned; surprisingly they contain much that has been overlooked. These include the books by Lother Machtan (2001), David Lewis (2003), and Traudl Junge (2003). It seems that, as John Lukacs has written, despite the existence of a vast Hitler industry, 'We are not yet finished with Hitler' (Lukacs, 1997, p. 51). To begin is to start with his parents.

Hitler's father Alois came from a family with a history of mental and physical problems (Waite, 1977). Born in 1837, he became a government official who enjoyed a certain success in his career and earned well, but he seems to have been an ugly brute in his personal life. He had a quick and violent temper, was promiscuous, faithless and a tyrant in the home. He fathered a child in his 30s to a woman whose last name is unknown. Then to advance his career, in 1873 he married Anna, the unattractive daughter of a senior government inspector and 14 years older than he. While married to her he carried on with his young housemaid Fanni, 24 years his junior, as well as with another young woman who also worked as a maid, and with his 16-year-old niece Klara. His infidelity was discovered and Anna separated from him in 1880, so he took up permanently with Fanni. In 1882 he fathered a boy, Alois Jr., to Fanni, and the following year, after Anna succumbed to tuberculosis, Alois married Fanni. Two months later she gave birth to a daughter, Angela, by him. However, Fanni developed tuberculosis and Klara was brought back to help and soon became his mistress. After Fanni died in 1884, Alois received a papal dispensation from his marriage to her, and when he married Klara in 1885 he was 48 and she was 25 and five months pregnant. After the wedding, husband and wife continued to call each other uncle and niece.

Klara took on the role of raising Alois's two children by Fanni and soon gave birth to Gustav. A year later in 1886 Ida was born, and in 1887 Otto arrived. However, all three of Klara's children soon succumbed to diphtheria; Gustav in 1887, Ida in 1888, and Otto just a few days after his birth. The effect of these deaths on Klara can only be imagined. Into this distraught emotional atmosphere, in 1889, Adolf was born.

Adolf was a sickly baby, which would have further aroused Klara's anxieties in the extreme. However for his first five years of life until 1894 when his brother Edmund was born, Adolf was his mother's only offspring (although not her only responsibility, for Alois Jr. was seven when Adolf was born, and Angela was six). As 'the only little boy I have left' (Waite, 1977, p. 206), Adolf was overindulged by his mother, who told him he was 'special', her *Lieblingskind* (darling child). Their family doctor, Eduard Bloch, later testified that in his long practice, 'I have never witnessed a closer attachment' (Waite, 1977, p. 169). This was also a time when the father was absent from the home, as his work took him away for extended periods. Hence in an emotional climate of traumatic loss, Adolf and his mother formed a cloyingly close bond in the frequent absence of a brutal, tyrannical father. In 1896 his sister Paula, his last sibling, arrived.

Although Alois was not quite, as Adolf later described him, the village drunk (he actually recalled his father as a drunken, sadistic degenerate), he was a heavy drinker and a very heavy smoker; Adolf remained revolted by smoking all his life. Alois physically abused Klara and frequently beat all his children. Rather than call Adolf to come to him he preferred to place two fingers in his mouth and dog-whistle as he did for the family pet, which he also beat mercilessly. Adolf remained ambivalent about his father all his life.

Throughout Adolf's childhood the family moved several times, and he went to three primary schools in three years. He was a good student at first, receiving good marks in the undemanding country schools he attended, although when his father was away Adolf manipulated his mother with tantrums, feigning illnesses to avoid school. Then in 1895, when Adolf was six, Alois retired at 58 after 40 years in the government service, and the family moved once more to a new home, a small acreage where Alois kept bees. Hence, in two years Adolf experienced a paradise lost when first his younger brother Edmund arrived in 1894, and then his father retired and was present in the home much of the time, and then in 1896 Paula was born.

Things must have been pretty grim because a year after Alois retired, Alois Jr., aged 14 and unable to endure living at home anymore, ran away to Vienna.

In 1897, Alois sold up and moved the family once again; he had found his small holding a burden and increasingly vented his frustrations upon Adolf. The first house they moved to was next to a blacksmith's shop, where the noise and smell gave Adolf a lifelong loathing of horses. Then Alois sold up again and settled in a cottage on half an acre not far from Linz. In 1899, Adolf started attending the Realschule, a kind of technical school, in Linz, and soon developed into a poor student, eventually repeating a grade. He hated this school, and it did not like him much either; his first end-of-term report described his moral conduct as 'adequate' and his application to his studies as 'erratic'.

Then in 1900 Adolf's younger brother Edmund died of measles. Again one can imagine Klara's reaction. However, something strange occurred in that neither parent attended the funeral, although Adolf did. It seems that Alois had fallen out with the local priest and refused to go into the church. He also refused to allow Klara to attend, taking her to spend the day at Linz; apparently she dared not disobey. Adolf stood alone in a driving snowstorm and watched the small coffin being lowered into a plot for which his parents never supplied a marker.

Adolf's personality changed after this; he became increasingly rebellious, which in turn led to even more frequent beatings from his father, but now he was determined not to buckle. There exists a perceptive account of Adolf's behaviour during this time, given by his former principal teacher Eduard Huemer at Hitler's 'Beerhall Putsch' trial in 1924, which suggests a bright but seriously disturbed child:

> Hitler was gifted, one-sided, uncontrolled, and was known to be stubborn, inconsiderate, righteous and irate; it was difficult for him to fit into the school milieu. He also was not diligent, because otherwise, with his talent, he would have been more successful; he could draw well, could do well in the sciences, but his desire to work always disappeared rapidly. Instructions and admonitions were received with undisguised irritation; from his schoolmates he

> demanded unconditional submission, (he) liked the
> Führer role ... (Cited in Redlich, 1998, p. 16)

Finally in 1903 Alois died, at the age of 65. Adolf, 14 at the time, reportedly wept as the coffin was lowered into the grave, but it is unlikely that he wept for the loss of a loving father, and in later years he would tell stories of how he hated his father (Waite, 1977). But at least the beatings stopped, although without his father's constant nagging his schoolwork declined even further. In 1904 he left the Realschule, probably because his mother and teachers thought he might do better elsewhere, and attended a boarding school in Steyr, some 25 miles distant, where his schoolwork did improve slightly. However, he left this school the following year after persuading Klara to let him stay at home to prepare for the entrance exam at the Vienna Academy of Fine Arts. Klara moved the family into a two-room apartment in Linz, where she and Paula slept in the lounge while Adolf had the bedroom.

Now, with money from the sale of the cottage and from Alois's pension, the family was comfortably off, and Klara was able to indulge Adolf with fine clothes and money to attend the opera. It was at this time that he began to say 'I shall become a great artist'. As a 15-year-old he spent hours designing buildings that he was absolutely certain he would one day construct, composing poems and an opera, painting unfinished masterpieces and — with utter conviction — dreaming of his future heroic artistic achievements. He seems to have lived in a grandiose fantasy world of art and success, a world of his imagination where painful reality did not intrude. At age 16 he became convinced that he could win the lottery, and he bought a ticket in one of his many 'great decisions' of this time. So sure was he that he would win that he planned in detail how he would spend the money, where he would buy his apartment, how he would furnish it and how he would create a great artistic salon that would draw in Vienna's cultural elite to hear him expound on aesthetics and give poetry readings. When he failed to win he raged against the lottery and the state. His teenage friend, August Kubizek, perceptively recalled that Adolf took the incident so badly because he felt that 'he had been deserted by his willpower' (Waite, 1977, p. 212).

What was the psychological reality of Adolf's childhood that led to him living in such a world of fantasy? Scholars have identified a curious passage from *Mein Kampf*, Hitler's later promotional autobiography written while he was serving time in prison, as being a thinly disguised biographical memoir that indicates the emotional tone of his home life. In this passage Hitler was describing the harsh social conditions that prevailed in Germany, and the life of a hypothetical child of a 'worker' in a dysfunctional family. Hitler wrote:

> Let us imagine the following: In a basement apartment of two stuffy rooms lives a worker's family ... among the five children there is a boy, let us say, of three. This is the age at which a child becomes conscious of his first impressions ... The smallness and overcrowding of the rooms do not create favourable conditions. Quarrelling and nagging often arise because of this. In such circumstances people do not live with one another but push down on top of one another. Every argument ... leads to a never-ending disgusting quarrel ... when the parents fight almost daily their brutality leaves nothing to the imagination; then the results of such visual education must slowly but inevitably become apparent to the little ones ... especially when the mutual differences express themselves in the form of brutal attacks on the part of the father towards the mother or to assaults due to drunkenness. The poor little boy at the age of six senses things which would make even a grown-up shudder. Morally infected ... the young citizen wanders off to elementary school ... The three-year-old has now become a youth of fifteen who [has been dismissed from school and] despises all authority ... Now he loiters about and God only knows when he comes home ... (Hitler, 1939, pp. 42–44)

Historian Waite (1977) notes ten similarities between this passage and Adolf's early life, and suggests that it probably indicates a primal scene in which his father raped his mother. Psychiatrist Redlich agrees — and Redlich does not agree with much in Waite (Redlich, 1998) — parallels with the Oedipal drama are obvious. It is impossible to be specific as to the meanings of the details of this passage, but Hitler gave another clue when, in the midst of further discussing deplorable scenes of domestic violence, he wrote: 'I witnessed all this personally in hundreds of scenes ... with both disgust and indignation' (Hitler, 1939, p. 38). As both Waite and Redlich point out, Hitler was no social worker so he simply could not have witnessed this hundreds of times outside of his own family. A cautious yet compelling interpretation is that this passage describes some important themes and events from Adolf's own childhood.

After his father's death Adolf lived a life that Alois would have utterly disapproved of. He counter-identified with his father by becoming first an effete dandy, and then by running off to Vienna to become a dropout bohemian artist. It is widely believed that Hitler was uneducated and uncultured but this is not quite true. He lacked formal qualifications because he did not finish school, but he read voraciously all his life and he spent many hours in libraries. As a youth he visited the opera frequently, and later he attended concerts whenever he could; Beethoven and Weber were favourites, but he also enjoyed the Jewish composers Mendelssohn and Mahler; he especially loved Wagner, identifying with Rienzi and Lohengrin. He loved theatre and was well-informed about painting and architecture. His weakness was that he was an autodidact, and thus his knowledge was unrefined, having gaps and prejudices that he was unaware of, but as Redlich notes, he had 'a strong synthesising mind' (Redlich, 1998, p. 177).

It is also often claimed that the many propaganda tracts and cheap novels that he read at this time account for his crude political ideas, his social Darwinism and his 'soil and blood' theorising. In fact, he also enjoyed Defoe, Cervantes and Twain, and his political ideas were shared by many who were much better educated than he was, most notably Zionist revisionists such as the notorious Stern Gang ('German blood for German soil, Jewish blood for Jewish soil'), of whom one member, Yitzak Shamir, even went on to become Prime Minister of Israel many years later (Brenner, 2007).

Then in 1907 Adolf's mother died from breast cancer. The circumstances of Klara's death have been widely debated. It has been argued that she died from medical malpractice by Dr Bloch, a Jew, and that this explains Hitler's anti-Semitism, but Redlich refutes this convincingly. Bloch seems to have been a model of compassion. Nevertheless Rudolph Binion (1991), author of this suggestion, has stated that, 'Consciously or unconsciously the bereaved always blame the doctor for the patient's death' (Lewis, 2003, p. 66). The coincidence of a Jewish physician treating the dying mother of the greatest anti-Semite in history should not be overlooked, but nor should it be overstated. In fact Hitler told Bloch, 'I shall be grateful to you forever', and he was, sending Bloch hand-painted thank-you cards years later. Probably all this coincidence really signifies is that Hitler was capable of distinguishing between compassionate individual Jews and their individual acts of decency, and what he believed was the scourge of the Jewish race as a whole (Redlich, 1998). In World War I he was awarded the Iron Cross by a Jewish army officer, and he seamlessly accommodated this to his malevolent script. The death of his mother did not make Hitler into an anti-Semite.

As she died, Adolf nursed his mother with devotion. He was devastated by her death; Bloch later commented that, 'In my entire career I have never seen anyone so prostrated with grief' (Lewis, 2003, p. 66). Many who knew of the relationship believed that Adolf's love for his mother 'verged on the pathological' (Waite, 1977, p. 169), and his friend Kubizek later spoke of Adolf's 'unique spiritual harmony' with his mother (Machtan, 2001, p. 42). Some understanding of the nature of their relationship is provided by an episode that occurred years later, when Hitler visited an art gallery with his friend Ernst Hanfstaengl and saw a painting of Medusa by Franz von Stuck. Hitler became mesmerised by the picture and commented to his friend, 'Those eyes, Hanfstaengl! Those eyes! They are the eyes of my mother!' This picture, and an 1895 photograph of Klara (the year of Edmund's birth, when Adolf was five), are reproduced in Waite's book. In the photograph Klara looks nothing like the Medusa picture, which has haunted, crazed eyes. The conclusion has to be that this was how Adolf experienced his mother. Whatever else he felt for her, and despite how he described his feelings, he experienced his mother as being what others would consider insane, although he would never have put it that way.

The next turning point in Adolf's life was his failure to gain admission to the Vienna Academy of Fine Arts. This actually occurred in 1907, a short time before his mother died, but because she was in the last stages of dying he kept it from her. It was a major blow to him, not only because he had staked his future on being a great artist, but also because he had convinced himself of the inevitability of success. He applied again the following year and was rejected once more. Ominously, four of the seven examiners were Jewish; he later claimed to have written a letter to the Director of the Academy that ended with the threat: 'For this the Jews will pay'.

Soon after this Adolf's friendship with August Kubizek deteriorated. It had been on the slide for some time because of Adolf's fiery outbursts and temper tantrums that had led August to suspect that the death of Adolf's mother had unhinged him. When Kubizek graduated from the conservatory with high honours, Adolf was there to congratulate him, but on returning from a stay in Vienna, August found that Adolf had moved and left no forwarding address. It was the end of their friendship. In the face of his own failures, Adolf could not endure August's success.

The loss of his mother and his artistic failure left him adrift in life, depressed and without prospects. He had a little money from his mother's estate, some of which he slyly allowed to be paid to him under the false pretence that he was a student, thus depriving his sister Paula of her full share of the inheritance, although he agreed to stop this after being threatened with court action. His aunt Johanna also left him some money when she died, but these sources of revenue soon dried up, leaving him impoverished. The self-belief that had sustained him now crumbled, and his next few years were very painful. He drifted through cheap lodgings, working as a labourer, accepting charity, living from soup kitchens and sleeping on park benches when the weather permitted. Poverty transformed him into an unshaven, dirty derelict. His finances improved a little when an associate advised him to paint scenes of Vienna to sell to tourists.

Adolf held himself together during this time by the use of rituals and compulsions that soothed his disillusionment and kept his hopes alive. He had a compulsive need to talk, to make grandiose speeches acting out the role of Messiah, and it seems that he lived through these speeches. He had begun speechifying when young; even at school he had made speeches and he had been considered odd for doing this. Kubizek later recalled one such speech, made when Adolf was 17, that he found profoundly moving:

> Never before and never again have I heard Adolf Hitler speak as he did in that hour as we stood there alone under the stars as though we were the only creatures in the world … It was as if another being spoke out of his body, and moved him as much as it moved me. It wasn't at all a case of a speaker being carried away with his own words. On the contrary, I rather felt as though he himself listened with astonishment and emotion to what burst forth from him with elementary force … it was a state of complete ecstasy and rapture … He spoke of a special mission that one day would be entrusted to him. (Waite, 1977, pp. 213–4)

In his adversity after his mother's death and his failed art school applications, Adolf continued to live out his Messiah fantasy by making speeches to anyone who would listen. Often he made them alone, in what therapists nowadays call 'self-soothing'. He was able to lose himself in a role, and he had a powerful imagination that enabled him to visualise himself in grandiose states (as a defence against his fallen state), and to act out climactic scenes in a quasi-operatic, heroic manner; much later at mass rallies he modelled his delivery on Wagner's operas. He raged against cruel fate and social injustice, planning and rehearsing in his imagination his as-yet-unknown great destiny, his vivid language lifting his mood to transcendent heights. He cried easily and was emotionally labile. He repeated and refined his stories, becoming a great actor, but only in his chosen role.

He also became a master of secrecy and a brilliant liar (children from abusive homes have to become skilled liars to survive). He was not a complete loner but neither did he really have close friends. His relationships were all one-way, on his terms, and he was hypersensitive to slights. During this period of adversity he was forced to develop his fantasy of being 'special' to greater heights just to stay sane, and in so doing he developed his superb rhetorical and theatrical talents to ever-higher levels. And there was one other disturbing issue in his life.

Much of Hitler's life and behaviour have been a mystery for biographers, and they have resorted to various stratagems to decipher his motivations. Hitler's lack of a personal life, his curious relationship with Eva Braun, and the extremism and ruthlessness of his hatreds seem uncanny and make him appear demonic. Yet Ian Kershaw, whose massive two-volume biography has been described as 'definitive' (Kershaw, 2000, 2001), ended up judging him as 'a man without characteristics ... take away what is political about him and there's little or nothing left'. [2] One comes away from such books with a sense of something missing.

In 2001, Lothar Machtan's *The Hidden Hitler* appeared, and much of the mystery fell into place. Machtan, a German historian, seems to have done what no one else has been able to do; he has identified and described the hidden dimension of Hitler's personal life. Machtan's thesis, exhaustively researched and cogently argued, is simply that Hitler was homosexual, that before he became a politician he was active in the homosexual subculture, that he suppressed his homosexuality while in the public eye, and that his sexual orientation explains several significant yet hitherto mysterious strands of Hitler's behaviour, such as his relationship with Eva Braun. To be fair to other researchers, some have come very close to identifying Hitler's homosexuality — Redlich diagnosed him as a latent homosexual — but none have assembled such a mass of data to make the case, although again Redlich comes close, citing the friendly witness H.S. Zeigler as saying that 'Hitler could not look at male ballet dancers and had to avert his eyes' (Redlich, 1998, pp. 282–283).

In his book, Machtan describes Hitler's adolescent relationship with his friend August Kubizek, his wanderings through the Viennese homosexual milieu, his liaisons with various homosexual friends over the years, his behaviour in the army where he twice received the highest award for bravery but was not promoted further than Corporal, and his repression and assassination of those who might compromise him, including his comrade and rival Ernst Rohm, also a homosexual and the leader of the Brownshirts. Machtan's book is very persuasive, yet tantalisingly incomplete. He discusses Maria Reiter, Eva Braun and Geli Raubal, some of Hitler's closest female associates, and shows that it is unlikely that Hitler ever had sexual relations with them (despite Reiter's suggestion to the contrary), but he does not mention others such as Inge Ley, Renate Mueller or Suzi Liptauer. Gay reviewers of the book have rightly been at pains to defend against any suggestion that homosexuality may incline towards antisocial behaviours, but this distracts from what surely has to be the main conclusion: while it was not Hitler's homosexuality that caused him to murder, nevertheless society's historical persecution of homosexuals may have had much to do with it, as per the Auden quote that heads this chapter.

As Alan Downs (2005) has argued, much gay behaviour is fuelled by shame and rage; the shame of growing up gay in a straight world, and rage at being stigmatised and feeling worthless. If channelled positively this rage can become an urge to excel and to prove one's worth, to transcend one's tormenters, and indeed, homosexuals are disproportionately represented in the arts, media, hospitality and entertainment industries. But a darker response may be a destructive anger and a burning desire for revenge. For such persons, hating is a kind of therapy; it holds them together emotionally and gives them direction and an explanation of painful life events. They need enemies, and almost anyone will do (obviously this scenario is by no means confined to homosexuals — anyone sufficiently stigmatised may develop in this manner). Hitler seems to have had all this and much more burning within him during his adolescence and youth. He probably felt doubly alienated, first as a homosexual and later by the rejection of his applications to art school. Already a misfit socially, mourning the loss of his mother and brewing with anger from his father's abuse and his artistic failure, resentment at his homosexual plight may have turned him into a walking time-bomb.

Adolf was rescued by the outbreak of World War I. It is difficult today to imagine a bohemian patriot; since the 1950s artists and bohemians have been political dissidents and pacifists. Hitler was a dissident all right, and a genuine bohemian, and he had left Austria to live in Germany at least in part to avoid military service (but also because Munich was at that time something of a homosexual El Dorado). But when Germany entered World War I, even such unlikely figures as Sigmund Freud became enthusiastically patriotic (Freud later admired Mussolini and sent him a copy of one of his books inscribed, 'To the hero of our culture, from an old man, Sigmund Freud' (Redlich, 1998, p. 138). The war was only expected to last for a few weeks or months and no one could foresee the carnage that lay ahead. Perhaps the most one can say of that curious (to us) confluence of attitudes that enabled highly unconventional people to be fervently patriotic is, first, that everyone felt a much closer sense of community at that time, and second, that the thought of war did not evoke quite the same feelings of horror then that it does today. War was a fact of life and an opportunity for young men to escape from their humdrum lives for a little adventure and advancement. Military men were still heroes; many operas included a dashing lieutenant and there was almost no criticism of the excesses of the male role. The bohemian milieu that Hitler lived in certainly disdained convention and authority, but it was quite natural for Hitler to be a radical critic of his society while also being eager to defend it by force of arms, especially as army life offered him a job and prospects.

Hitler joined the List Regiment and served at the front, but not as a front-line soldier. It seems he reached the front in the last week of October 1914, and was promoted to Gefreiter or Lance-Corporal in the first week of November, but he was made a Meldeganger or dispatch rider, moving around behind the lines carrying orders from headquarters to the units and back again, and he remained in this position for the rest of the war. This was still not a completely safe place to be, of course, and he had several narrow escapes and was slightly wounded twice, but he was also considered to lead a charmed life. He came to believe he was protected by an 'inner voice' that pre-served him for some great destiny (Lewis, 2003, p. 112). A comrade who served with Hitler has described his war service thus:

> Hitler never had anything to do with guns from the time
> he joined us at the front as a regimental orderly. He was
> never anything other than a runner based behind the lines

at regimental headquarters. Every two or three days he would have to deliver a message; the rest of the time he spent 'in back' painting, talking politics and having altercations. He was soon nicknamed 'Crazy Adolf' ... He often flew into a rage when contradicted, throwing himself on the ground and frothing at the mouth ... The List Regiment's battalion adjutant was Lieutenant Gutmann, a Jewish typewriter manufacturer from Nuremberg, whom Hitler made up to whenever he wanted preferential treatment of some kind. It was Lieutenant Gutmann who got him his Iron Cross Second Class at Christmas 1914. That was in Bezaillere ... Colonel Engelhardt of the List Regiment was wounded in this engagement. When he was carried to the rear Hitler and Bachmann tended him behind the lines. Hitler contrived to make a big fuss out of this exploit of his, so he managed to gain Lieutenant Gutmann's backing ...

Hitler could never forebear to deliver inflammatory political speeches to his comrades ... (he) struck me as a book with a thousand pages ... two-faced ... hypocrisy personified. One of his faces was that of the self-important busybody he impersonated to his superiors ... Hitler's other face was that of a secret sinister criminal. His whole attitude was that of a ruthless person who knows how to wrap himself in a halo. He has always, ever since I've known him, been ... a great actor. Not a word he uttered could be trusted. (Machtan, 2001, pp. 67–70)

The most curious thing about Hitler's military service is that although he was twice awarded the Iron Cross (he won it in both classes and it was rare for anyone other than an officer to be awarded it First Class, although towards the end of World War I, Iron Crosses were being handed out more freely in order to raise morale), yet he was never promoted past the rank of corporal. The official reason was that he lacked leadership qualities, a revealing claim with two likely strands: either he was recognised as a homosexual or a social misfit or both, and thus deemed unfit.

In mid-October 1918, Hitler's unit experienced a gas attack that left him temporarily blinded. It probably was not mustard gas as he claimed, but the milderWhite Star gas, although he may have sincerely believed it to be mustard (Lewis, 2003). Rather than being treated at a medical unit in the field he was taken 600 miles across Germany to Pasewalk, a town with seven medical hospitals set up by the authorities to treat injured soldiers. The facility that Hitler was admitted to specialised in nervous disorders, hence it is almost certain that Hitler's injury was not physical but emotional; he had collapsed in the heat of battle and now suffered hysterical blindness, a fairly common occurrence. However, he probably believed that his ailment was physical rather than psychological (he considered psychiatry to be 'Jewish medicine').

What happened next is uncertain but very important. Two versions exist that are quite compatible with each other but both rest on sketchy supporting evidence. In the first version, given several times over the years by Hitler himself, as he was lying in his hospital bed, he received the news that Germany had surrendered. He reacted with anger, grief and despair, and in this time of extremity he received a supernatural visitation; he heard voices calling to him telling him that it was his destiny to rescue the Motherland from the Jews who had violated her. This led to him making 'the most decisive decision of my life … I had resolved to become a politician'. From then on he believed that he was being guided and protected by fate, saying: 'I go the way that Providence dictates for me, with all the assurance of a sleepwalker' (Waite, 1977, pp. 30–31, 246). In sum, Hitler had a transformative, quasi-mystical experience.

In the other version, promoted in the book *The Man Who Invented Hitler* by psychologist David Lewis (2003), Hitler was treated at Pasewalk Hospital by Dr Edmund Forster, who had developed an experimental method for the treatment of 'shell-shock' cases. What this involved was to make an appeal to the sufferer's vanities and core beliefs about himself. Forster would have made a thorough physical assessment of Hitler and recognised his problem as psychological rather than physical (there was nothing wrong with Hitler's eyes). He would not have disclosed this, for to do so would alienate his patient. Forster would also have noted Hitler's core values and self-concept, especially his self-concept as a man of destiny, and his need to feel strong. He would then have suggested to Hitler that unfortunately, most persons suffering such eye damage usually did not recover. This would have been partly true insofar as cases of hysterical blindness do not usually recover when treated by standard medical treatments such as drugs, nor by psychotherapy. However, Forster would have added that, on very rare occasions, certain highly exceptional individuals possessing great willpower were able, in the fullness of time, to heal their injuries by their sheer willpower alone, as if their mind somehow managed to stimulate their body's natural healing processes to restore the damage done, and to enable them to make a full recovery.

If something like this treatment actually happened, and it may well have, the result would have been to stimulate Hitler to 'heal' himself, pitting his need to avoid reality through hysterical blindness, against his need to feel special. The greater need won out; also, the war was over now so Hitler could safely reclaim his health without fear of being sent back into action. The result of this would be to confirm to Hitler that he was indeed special, an exceptional person able to perform miracles, guided and protected by fate to fulfil a unique destiny.

Both these scenarios are compatible with each other. If Hitler was given such a 'narrative intervention' or 'paradoxical instruction' (as such treatments are sometimes referred to nowadays) by Forster, he may well have responded to it by coming up with a mystical experience some time later. It is not uncommon for some therapeutic insight (which may or may not be true) to be followed by an intense 'aha' moment of great emotion later in a patient's week. This, in turn, may result in significant behaviour change, sometimes even personal transformation, further down the track. There is not really anything very unusual or controversial about such a treatment, although it is tricky and unconventional, but it is still well within the capabilities of very creative therapists such as Milton Erickson (Bandler & Grinder, 1979; Haley, 1973). We also know that mystical experiences are largely self-healing attempts by persons who need to transform themselves and their lives. Such experiences perform a large-scale 'gestalt switch' to one's belief system and world-view, which in turn provides a new self-image and a role that solves painful problems from the past, while opening up opportunities for the future (Batson & Ventis, 1982; Oakes, 1997, pp. 98–113. Highly creative individuals (and Hitler was creative) are well capable of such transformative experiences. As he wrote much later, 'Nothing is impossible; one can do everything if one has the necessary will'.[3]

The following year, 1919, was a year of recovery and exploration for Hitler. Terrified at the prospect of drifting back to the poverty and squalor of his Vienna days, he did the rounds of demobilised soldiers' organisations and the political underground. He managed to get himself a position as an undercover informer spying on rogue political elements. It was also the year that his anti-Semitism crystallised.

Redlich has described five stages in the development of Hitler's anti-Semitism in which, during his youth and much of his adolescence, he was no more or less anti-Semitic than his neighbours. However, after moving to Vienna, which was more stridently anti-Semitic, and after being twice rejected by the Vienna Academy of Fine Arts, his prejudices intensified; he was vulnerable at this time because of the recent death of his mother. At this point he was probably anti-Semitic in the way that his hero the composer Richard Wagner had been; Wagner was virulently anti-Semitic, but he also had several Jewish friends; it may be mistaken to see the prejudices of such men in overly-simplistic terms (Brenner, 2005).

The third stage occurred after World War I when he discovered that Jews were leading the Communist Raterepublik. His concern with social justice had earned him the army nickname of Red Hitler, and after the war he had hoped to join the communists, but he received no advancement there, perceiving this as another rejection by Jews. He was also devouring all manner of political tracts at this time, most of which were racist, and again he was vulnerable through being impoverished and depressed. He came to see world Jewry, not merely individual Jews such as the kindly Dr Bloch, as the great enemy, believing that the Jews had originated both Bolshevism and capitalist democracy. During this period he would have also recognised the value of using the Jewish scapegoat to arouse the masses. In his fragmented mind he could come to believe in anything that had such utility.

The fourth and fifth stages took place after seizing power and during World War II, and both involve the execution of his anti-Semitism rather than any elaboration of it. Perversely, it now seems likely that the 'Jews' who suffered most under Hitler were not Semites at all. They were Eastern Jews who were probably descendents of the Khazars, a Turkic group who came from the area between the Volga and the Caucasus, and who had converted to Judaism in the Middle Ages, but had subsequently migrated west (Haffner, 1979).

Hitler's anti-Semitism has sometimes been thought mysterious insofar as it seems to have had no specific cause. His rejection by the art school may have made him bitter, but his anti-Semitism was of another magnitude. Most likely he developed the clinical condition known as an 'overvalued idea' that was first described by psychiatrist Carl Wernicke.[4] An overvalued idea is not an obsession — some absurdity unique to the individual; it may be shared by many people to a lesser degree. Its function is to cohere an otherwise unstable personality around a single focus, a 'monomania' that gives destructive passions a positive rationale and an outlet. This allows a hostile person to express their hate while still seeing themselves as loving, and it enables him to present himself to others as compassionate and idealistic. The charismatic personality struggles to manage his passions, most especially his anger towards a world that refuses to acknowledge his narcissistic claims. He is at his best when he can channel his hatred into a just cause, casting himself as a loving leader who hates evil. Any evil will do of course, provided it coheres and stabilises. Hitler magnified his anti-Semitism when he finally had the opportunity to raise himself up, and when he most needed a stabilising force in his psyche, one that would be acceptable to others. As he advanced towards leadership, anti-Semitism gave him a precarious balance.

In 1919, Hitler was 30 years old and Germany was teetering on the edge of social collapse. He had learned much in the war: how to negotiate his way through the mainstream (the ones who saw him as a misfit), how to survive danger (actually, not new, rather an extension of what he learned from his father's beatings), how to nurture opportunities and exploit them, how to pace himself and his ambitions ('with all the assurance of a sleepwalker') and how to milk relationships with seniors. He probably also realised that he functioned much better when those around him were in crisis, than when things were less stressful. He now glimpsed a destiny; that is, he had a cause to wrap himself in, to cohere his turbulent personality and the lift in his self-esteem from his two Iron Crosses must have been great. Now that all of Germany was as lost and confused as he had been before the war, he knew well how they felt. He was on familiar emotional territory while they floundered in anxiety and the strangeness of poverty and lack of prospects, as he had once done. He would have seemed like one of the strong ones now, someone who was not daunted by adversity, someone with proven credentials of courage and resourcefulness, a potential Führer even. His anti-Semitism seemingly explained everything. In the chaos of defeated Germany he soared like the chosen one.

Hitler next joined a small extremist political party, the Nazi Party, and became a leader and soon a public speaker. He was recognised for his speaking ability immediately, but his breakthrough came on February 24, 1920, when he addressed a large rally. As outlined above, young Adolf had taken up speechifying as a ritualistic self-soothing strategy to cope with stress and to lift himself out of depression; while in school he had been noticed by a teacher 'holding dialogue with trees stirring in the wind', and others had observed him going out alone at night to make speeches upon a hill (Waite, 1977, p. 188). As he moved into public speaking he had a few early failures (Lewis, 2003, p. 187), but his years of performing before an imaginary audience soon bore fruit as he swept crowds away with the power and passion of his words.

Perhaps too much has been made of Hitler's rhetorical ability; it was not the sole source of his success (he was also a brilliant organiser). Many people unsympathetic to his politics came away unmoved. His voice was quite guttural and it did not sound so impressive on radio. His words lacked 'the humane note, the magnanimity of a cultivated human being' (Waite, 1977, p. 69), and his speeches culminated in orgiastic expressions of hate ('If a people is to become free it needs pride and willpower, defiance and hate, hate, and once again hate'; Baynes, 1942, p. 44). Educated people ignored him and Christians dismissed him, mistakes that would cost them dearly.

But Hitler's speaking ability was the base that supported his entire mission (he said so himself). He had that uncanny knack of charismatic personalities of being able to read the unconscious hopes and fears of others, and to articulate their needs and truths before they themselves could, thereby seeming to be a visionary, to be one step ahead of his audience. The suffering that poor Germans were facing after World War I was like the suffering he had endured all his life so he knew it from the inside; he could express hurt and rage with a familiar intimacy and authority. He had the credentials of his two Iron Crosses to show them that he had suffered too, and thus to pose as a prophet of the great salvation to come. By whipping himself and his audience up into an ecstatic frenzy he was giving them a glimpse of Nirvana, taking them into a liminal space beyond roles and statuses to taste the new sense of brotherhood that he embodied, in which the old is sloughed off and a new heaven on earth is touched and felt. His self-intoxification became theirs, turning their fear and pain into hopeful rapture.

For these reasons Redlich, at the end of his long and tightly constructed biography of Hitler, correctly describes him as a prophet, albeit a destructive one. He was much more than just a politician or a revolutionary or a demagogue or a dictator. He evoked people's ultimate concerns in an ecstatic, shamanic way, seeming to enter a personal relationship with each of them, to speak to each as if from inside their own hearts and minds, and to reveal them to themselves (see chapter 10 on Max Weber for a discussion of this). Hitler was something of a mystic, following his inner voice with absolute certainty of his calling, drawing sustenance, inspiration and (he believed) protection from this divine guidance. When, in 1945, he narrowly survived an assassination plot, rather than suspecting that providence may have failed him, he saw his survival as proof of the guiding hand of fate (Junge, 2003), and he retained this conviction until the end. The lesson has to be that the psychology of shamans and prophets is equally harmonious with good and evil intentions, and that many such figures stand ever-ready on the fringe of society awaiting the right circumstances to emerge.

Hitler's political career is too well-known to recount here, but more can be said of his personality. Fundamental to Hitler's influence was the powerful impact of his eyes; almost every biographer has remarked upon it. Redlich says that his eyes could look cruel or kind 'mostly according to the viewer's interpretations' (Redlich, 1998, p. 265), suggesting that there was some kind of placebo at work, that perhaps Hitler's eyes were unremarkable and that people just interpreted them as remarkable. However, even those who knew him long before he became famous remarked on how young Adolf 'spoke with his eyes' (Waite, 1977, pp. 5–6), hence it is unlikely that expectations account for all of the impact. It is known that he practised making piercing eye contact before a mirror, that he made a point of meeting the eyes of troops on parade, and that he stared meaningfully at passing strangers, but his powerful impact was probably the sum of his eyes, his voice, his posture and gesture, and his active, expressive face.

That Hitler's eyes were identified as the focus of his charisma may be simply because the eyes are what observers tend to concentrate upon in another person. However, when someone is totally convinced of their 'truth' certain subliminal signals leak out, and are often expressed through the eyes. An unexpected absence of anxiety may be expressed as a soft stillness, an ease with conflict can be signalled by a comfortable arousal, an intense conviction and a careless confidence may be revealed by reassuring warmth. These may all be conveyed through the eyes, or at least, that is where we look, although such messages are also conveyed by posture, voice and facial expression. Thus the impact of Hitler's eyes probably derives from three sources: the total impact of all his para-linguistic behaviour (gesture and voice tone and so on), his inner emotional state of intense certitude, and to some degree the expectations of those he came into contact with.

Hitler also radiated a certain power, and again more was probably involved than just the expectations of those around him. During the war years, Admiral Donitz felt psychically drained by being around him, and Albert Speer also came away feeling exhausted; both avoided him, the better to 'preserve [their] own power of initiative ...' (Waite, 1977, pp. 455–456). The effect of his power during this time has been described by his secretary Traudl Junge:

> Sometimes I saw Hitler's advisors, generals and colleagues come away from talks with the Führer looking dismayed, chewing on thick cigars and brooding. I spoke to some of them later and though it often happened that they were stronger, wiser and more experienced than me, it often happened that they went to see the Führer armed with unimpeachable arguments and documentary evidence, absolutely determined to persuade him that an order was impossible or could not be carried out. But before they had finished he would begin talking and all their objections melted away, becoming pointless in the face of his theory. They knew it couldn't be right but they couldn't pin down the flaw in it. When they left him they felt despairing, crushed, with their former firm and absolute resolve badly shaken, as if they had been hypnotized. I think many of them tried to hold out against his influence but others felt exhausted and worn down, and then just let events take their course to the bitter end. (Junge, 2003, p. 109)

Again, part of this effect on others would have derived from his inner certitude, combined with the expectations of those who encountered him. But Hitler was also profoundly manipulative; his pioneering use of the media to project his message derived from his own creativity, and there was a chameleon-like quality to him that always came up with something dramatic and unexpected, so he probably manipulated others into dependent states before he went to work on them. He used his extraordinary memory to prepare for meetings with his senior military staff, bombarding them with questions about armaments, tonnages, casualties and military history in order to intimidate them into silence. He might query some aspect of an officer's career such as the name of the person who presented him with a particular medal, or the names of the wives of his junior officers, and then, when the officer could not recall such details he would announce them triumphantly, thereby proving that he knew more than they did and was thus better able to make decisions than they. He frequently claimed to be an expert on any discipline (Schweitzer, 1984, p. 82).

He was able to turn on the tears, to rant and rage, and then to become sweetly charming, languid and calm. Indeed, many who met him remarked upon a softness, almost a femininity about him in the way he walked, his gentle demeanour and his lulling words, until the hate-filled ranter emerged again (Machtan, 2001). It is important to understand just how soft and playful and even somewhat flexible he was when not in his Führer role. When observers report that he was wild and cruel and fanatical and rigid, their accounts are true only for those moments when Hitler was consciously and deliberately making such impressions (his constant acting again). But he was also able to find pleasure in simple things, and in warm homely exchanges. During the last years of the war he held nightly after-dinner gatherings in which witty banter and mundane gossip were exchanged, although his narcissism was never far beneath the surface and sometimes leaked through. Two further extracts from Junge illustrate this:

> In the tea house Eva told the Führer he shouldn't stoop like that. 'It comes of having such heavy keys in my trouser pocket,' he said. 'And I'm carrying a whole sackful of cares around with me.' But then he couldn't help making a joke of it. 'If I stoop I match you better. You wear high heels to make yourself taller, I stoop a little, so we go well together.' 'I'm not short,' she protested. 'I'm 1.63 metres, like Napoleon!' No one knew how tall Napoleon was, not even Hitler. 'What do you mean, Napoleon was 1.63 metres tall? How do you know?' 'Why, every educated person knows that,' she replied, and that evening when we were in the living room together after dinner she went to the bookcase and looked in the encyclopedia. But it didn't say anything about Napoleon's height. (Junge, 2003, p. 119)

> Hitler began quietly whistling a tune. Eva Braun said, 'You're not whistling that properly, it goes like this.' And she whistled the real tune. 'No, no, I'm right,' said the

Führer. 'I bet you I'm right,' she replied. 'You know I'll
never bet against you because I'll have to pay in any case,'
said Hitler. 'If I win I must be magnanimous and refuse to
take my winnings, and if she wins I have to pay her,' he
explained to the rest of us. 'Then let's play the record and
we'll see,' said Eva Braun. Albert Bormann was the adjutant on duty. He rose and put the record in question — I
forget what it was — on the gramophone. We all listened
hard and intently, and Eva Braun turned out to be right.
She was triumphant. 'Yes,' said Hitler. 'So you were right,
but the composer composed it wrong. If he'd been as
musical as me then he'd have composed my tune.' We all
laughed, but I do believe Hitler meant it seriously. (Junge,
2002, p. 80)

Because of the magnitude of his crimes, there is a tendency to see
Hitler as a caricature, as a one-dimensional psychopathic demagogue
with a comic quality à la Charlie Chaplin in the film *The Great
Dictator*. But malignant narcissism need not be like that. On the
surface such persons may be charming and socially adroit, indeed, the
most charming and socially adroit people one may ever meet. These
two extracts suggest that Hitler had a warm and gentle sense of
humour, could accept being wrong on occasions, and could even tolerate having the wool playfully pulled over his eyes, but it all
depended upon context. He was only play-acting the role of Eva
Braun's partner (to her as well, although her loved her in his way,
perhaps as one loves a faithful pet). What is endlessly astonishing
about him is the utter separation that existed between such mild and
homely behaviours as those described above, behaviours that seemingly show a warmth and mundane humanity that would be quite
unable to countenance harming another, and the pathological component of his personality that co-existed quite comfortably alongside
it. It is as if two utterly contradictory parts of his personality existed
independently side-by-side, so that the one never reflected upon or
mediated or interacted with the other.

Such 'fragmentation' as Kohut called it, or 'mental splitting' as it
is otherwise known, is stunning to observe in clinical settings, consti-
tuting a real 'split personality' (although not in the sense associated

with schizophrenia, which is really more like a broken or disorganised personality). What held Hitler's psyche together was an underlying narcissism that saw the entire world as merely an extension of his ego; that felt no need to reconcile contradictions in his personality to fit in with the world; and that was driven to megalomaniacal binges of hatred and destruction in order to avenge himself upon a world indifferent to his special claims. Yet he also developed — as most of us do — a quite culturally sensitive and socially aware sympathy that could enjoy homely games and gentle ribbing, so long as it came from inferiors. A more 'normal' person would be unable to tolerate the dissonance between the two parts of themselves, but such splitting had probably once been necessary for young Adolf to survive in his family of origin, and he took it for granted and thought little about it. Indeed, he prided himself on resisting change, on not compromising with reality and on nurturing his contradictions and idiosyncrasies regardless of what others might think. In a crazy family such as he grew up in, such contradictions would have been ignored, as they were again in wartime and in the equally disturbed post-World War I Germany. The traits that seem so bizarre to us in retrospect were once quite functional for him, and they became functional again when he found himself in another deeply disturbed milieu.

Perhaps the most striking thing about Hitler's 'down time', those hours when he was not formally on display and could relax, was that although he had a warm sense of humour that co-existed comfortably alongside his ability to wage war and oversee the holocaust, nevertheless, he could not commune, could not completely let go of his controls and be an equal exchanging simple banter, although it looked as if he could. Junge has also recalled how, after he had gone to bed, the atmosphere in his retreat lightened and people spontaneously laughed and had fun, as they had been unable or unwilling to do in his presence. Even when relaxing Hitler had to be in control, no matter how subtly, and even in his quietest moments he exuded a pressure that inhibited others. This is best illustrated by his critical attitude. As Kohut has argued, the self-esteem of charismatic personalities is supported by their habitual need to criticise others, to point up others' deficiencies in a way that may seem innocently corrective, yet that also serves to demonstrate their own presumed superiority (Kohut, 1976). The form this may range from mere gentle ribbing of friends and associates about the dangers of smoking or meat-eating or other vices (Hitler consistently did this; Junge, 2003), all the way to making war on 'inferior' peoples.

But what was Hitler really doing in these after-dinner soirees? Ostensibly he was relaxing with friends and associates after a hard day's work. He would stay up late with his staff to receive the reports of air-raids and battle movements, then unwind with pleasant company chosen by him. This was the nearest that Hitler ever got to routine, intimate family life. His conversations with his admirers were the first — and really the only — time in his life when he was able to indulge in normal social behaviour, and it is clear that he hungered for it. The lesson is, however, that Hitler could only allow himself such luxuries when totally dominant, when surrounded by a coterie of dependants, toadies and unworldly innocents like Junge, who were unable to challenge or even to comprehend him. Further, even at such moments he still could not really drop his defences and controls (they had long been habitual) and openly commune in the way that normal people do, and his companions felt this; only after he had gone to bed did they genuinely relax.

Why could Hitler not see that his companions were not really relaxed? Perhaps because he had never experienced genuinely relaxed company in his early life. Throughout his formative years his home life had been so tense that a low-grade tension felt normal to him, as did his need for dominance and control. Such compulsive self-control and deep insecurity are seldom transparent; they are usually only deduced retrospectively as explanations for striking behaviours. Hitler longed for free and open commensality but was too insecure to engage in it, achieving only a shadow of it in these contrived gatherings with his hand-picked guests. In one part of his mind he was asking the attractive young Frau Junge for another lump of sugar please, while enthusing over her wedding plans, but in another part he was waging all-out war and guiding the holocaust. In both parts he was powerfully manipulative and controlling. An indication of the depth and sophistication of his manipulativeness can be seen from the following account given by Junge. The scene is his bunker in May 1945, just after Hitler has received news of the Russian encirclement. He gives an order to his personal staff to abandon Berlin and fly south to safety, then:

> Eva Braun ... goes toward Hitler ... takes both his hands and says, smiling and in the comforting tones you might use to a sad child, 'But you know I shall stay with you. I'm not letting you send me away'. Then Hitler's eyes begin to shine from within, and he does something none of us, not even his closest friends and servants, have ever seen him do before: he kisses Eva Braun on the mouth while the officers stand outside waiting to be dismissed. I don't want to say it, but it comes out of its own accord; I don't want to stay here and I don't want to die, but I can't help it. 'I'm staying too,' I say. (Junge, 2003, p. 163)

Another curiosity commented upon by several commentators is the lack of growth and development in Hitler's personality (Haffner, 1979, p. 51). He was still saying and doing the same things at 55 that he was at 25, recycling the same ideas, following the same ideals, nurturing the same hatreds and suffering the same mood-swings. He seemed to remain ready for suicide for most of his adult life, certainly he spoke of it often, and he never achieved a close, equal, intimate relationship with any other. Nor did he develop a capacity for real friendship, although he loved animals and he allowed himself some soft feelings for Eva Braun (albeit he totally dominated her). Such lack of development is not uncommon among charismatic personalities, many of whom seem to prefer to find a niche of narcissistic supply and then to try to remain safely within it. It also indicates a lack of healing.

Lastly there is Hitler's tremendous destructiveness, which revealed itself in unexpected ways. Not only was he the architect of a massive war of conquest and the Final Solution, he also took some actions that seem not to make sense from any military viewpoint, and which can really only be understood as expressions of consuming hatred. For example, there was no need for Hitler to declare war on America when he did (although he had a rationale; Kershaw, 2007), and it enabled Roosevelt to abandon restraint and enter the European war. Similarly, in the last weeks of the war Hitler issued several orders, the effects of which would have been to destroy Germany itself, and to punish the surviving population for not supporting him enthusiastically enough. Fortunately, these orders were mostly ignored. This destructiveness of Hitler has led some, notably Clive James, to conclude that ultimately, if given his way and enough time, Hitler would have murdered everybody.

Was Hitler a psychopath? In a trivial sense, of course he was; he murdered millions without regret. But, as Redlich alludes, he does not fit well within the criteria of the antisocial personality disorder (Redlich, 1998). Before he became the greatest criminal of all, he did not repeatedly perform criminal acts, he was not involved in frequent physical fights, he did not show a reckless disregard for the safety of himself or others, and he was not consistently irresponsible in his work or financial obligations. Hitler's murderousness did not arise solely from some vicious hate-filled psychopathology, albeit he had that too. Rather, it came from pushing an ambitious doctrine to its limits, ruthlessly and remorselessly. Alexander Solzhenitsyn once considered why bloodthirsty Macbeth killed only a handful of people while Stalin murdered millions. His answer was that Macbeth had no ideology to push to its extreme. This was the nub of Hitler's criminality.

Commentators often suggest that Hitler's crimes are unthinkable, but in fact, Hitler's ideology and his inclinations are all too thinkable. Historically, genocide has been the ultimate way of resolving tribal conflicts once and for all. Most of us fantasise occasionally of ridding the world of evil (variously defined), but while this is thinkable it remains un-doable. For Hitler it was not. He was able to mobilise his fantasies with all their horrendous implications without inhibition because he had no genuine sense of involvement with others, no investment in them, and no internalisation of them within himself. Too fragmented to be anchored at his centre, and too narcissistic for connection, there was nothing holding him back (as in Yeats' memorable poem, 'things fall apart, the centre cannot hold' Yeats, 1996, p. 187). He was not close to any other; his relationships were opportunistic and his friendships existed in fantasy only. He was able to execute his political program without so much as a backward glance at the ghastly trail of death in his wake. In his early years as leader he had a few regrets about killing Ernst Rohm and one or two others, but he rationalised these as necessary acts that no effective leader could shrink from, one of the burdens of high office. He had not attained the kind of mature ego-development that enabled him to place the welfare of others above his own, or even on a par with it. He was so emotionally stunted that his ideology was more real to him than the flesh and blood of other human beings, even those he most cared for.

It must be added in passing that this stance of ideology over persons is a long-understood problem within mainstream Christianity, and no doubt of other religions too. In its simplest form it is mere fanaticism, the kind of extremism that blinds the fanatic to the sufferings that dogma can cause to others, but even most fanatics will baulk at some things and can be led to remorse. Hitler's fanaticism was propelled by such a severe impoverishment of spirit, such a fragmentation of his sense of self, that he had no feeling for proportion or balance in human affairs, which suggests that relationships existed purely as fantasy to him, just one more complicated aspect of his troublesome psyche that was always threatening to collapse or disintegrate.

Our understanding of psychopathy (or antisocial personality disorder) has come a long way in recent years, and it is now generally accepted that for a person to become a 'dangerous violent criminal', certain experiences must have happened to them in their formative years. Criminologist Lonnie Athens (1992) has called this process 'violentization', and has described it as comprising four stages: brutalisation, belligerency, violent performances and virulence. Without going too far into the details of Athens's theory, it is clear that young Adolf did experience brutalisation in all its subcategories of violent subjugation, personal horrification and violent coaching, a process that Athens advises leaves its victim 'deeply troubled and disturbed'. However, he seems not to have experienced the second stage of belligerency, nor the third stage of violent performances in the sense that Athens intends — violent fights. Perhaps Hitler's artistic side, or even his incipient homosexuality, inhibited him. Then again, he did experience World War I.

The final stage of virulence, that is, moving from a defensive to an offensive stance, may be relevant. Athens (1992) advises that in this stage the violence that had characterised the earlier stages and that had mostly been used as a defence, a method of intimidation based on the principle that the best form of defence is offence, comes to be seen as a tool for advancement in the world. It becomes a strategy for achieving success. Now the 'victim' seeks out conflict.

There is some (modest) similarity here with two aspects of Hitler's behaviour. First was his political philosophy that invoked crude social Darwinism, or 'might is right'. Second was his uncompromising use of hatred and violence as a means of advancement. As Haffner points out, Hitler really only had one strategy and that was to attack, again and again. In his political takeover of Germany there was little political strategy, and certainly no subtlety. Later in his military offensives there were no feints or strategic withdrawals, all he ever did was attack, this sole methodology being supplemented only by an exquisite sense of timing. When it became clear that he had overreached in Russia, he refused to regroup (that would have meant conceding error), but instead he relied upon efforts to bolster his troops with air-supplies and reserves. This blunt approach worked well with weak opponents, both in politics and war, but it failed utterly with equal or superior opponents (Haffner, 1979). In sum, while Hitler does not fit neatly into characterisations of violent dangerous criminals or psychopaths, and nor does he fit the description of individuals who commit large-scale evil (Waller, 2002), nevertheless his development probably involved some aspects of the violentisation process described by Athens.

The case of Hitler highlights the controversy over the diagnosis of personality disorder. Much of his adult behaviour suggests that he was mentally 'sick'. Indeed, his adult behaviour was so destructive that if we cannot call him 'sick' then we might wonder just who on earth does qualify? But knowing his background enables us to say that in fact he showed exceptional resourcefulness and adaptability just to survive his childhood. He seems to have been well-adjusted to the crazy environment that he grew up in, better perhaps than most other more 'normal' people might have been had they grown up in it.

Certainly he was not sick in the medical sense, as is a schizophrenic or mentally-handicapped person. The most we can say is that he failed to develop the characteristics we expect of normal adults, including having a conscience, empathy and a sense of social responsibility. But does such a failure to mature constitute sickness? This reveals the moral dimension of psychiatry, although it can equally be argued that, as we are social animals, then such a failure to develop does constitute a sickness (or at least a disability) in such a fundamentally social animal as man.

What conclusions can be reached about Hitler from the perspective of the theory of charisma proposed herein? It would seem that he embodied the kind of genius combined with the psychopathology to which William James referred. His psychological contradictions arose from the kind of fragmentation of his self that Heinz Kohut described. He was inclined by painful early experiences to develop the false self that James Masterson has outlined. This enabled him to feel worthwhile, and gave him a role. He was so severely neurotic that, in time, projecting this persona became the only way he could feel comfortable with himself. He had powerful reasons to hate, but his artistic temperament and perhaps also the moderating effect of his homosexuality might have enabled him to avoid degenerating into psychopathy, although in retrospect that would have been the lesser evil. As a young adult with a narcissistic mentality he soon found himself floundering, lost and alone, much more so than his peers with whom he felt few connections. But then came the time when his experiences resonated with the experiences of others. He had the talent to read the hopes and fears of his countrymen, and the inclination to recognise himself as a prophet, and the skills to lead a charismatic movement. When the chance came he seized destiny and attempted to reshape life closer to his heart's desire, not into what we might recognise as a benign utopia, but into something that reflected his disturbed origins. In pursuit of his vision he resorted to large-scale evil because he was not restrained by connectedness to others (the narcissistic mindset), or the fruits of connectedness: empathy and largesse of spirit.

The impression we are left with is of a very disturbed person who never connected with others, who never developed the kind of personal security that could enable him to admit when he was wrong, whose paranoia and homosexual anxiety was ever-present, who struggled to survive and only grasped at the role of politician because it seemed to offer a way forward, a way to end the struggle within himself and to see himself in a positive light. He was a fierce pretender who lived a false self all his life, glimpsing his higher calling only through the veil of his own hatreds. He was detached from all that gives a human life its meaning, but powerfully creative and intelligent (a perverse testament to the human spirit). He could not enjoy the womb-like satisfactions of maturity — adult love and commensality — because his personality was structured against these. Eventually his great quest failed, leaving him once more hurting and alone, but now with no further chance to redeem himself.

Perhaps the final words on Hitler ought to be his own, for they reveal the elusive insight into themselves that charismatic personalities occasionally show. In a statement to his doctors, made after taking power in 1933, he revealed, 'I suffer from tormenting self-deception' (Waite, 1979, p. 43). Years later, in one of his monologues he stated, 'My entire life has been nothing but a continuous persuasion' (Redlich, 1998, p. 303).

Endnotes

1 For an example of this kind of research see, Donald Dutton's (2007) *The Abusive Personality: Violence and Control in Intimate Relationships*. Although Dutton's book was published after Redlich's, nevertheless much of the material in it has been around for longer, and his earlier book *The Batterer: A Psychological Profile* (1997) appeared the same year as Redlich's biography of Hitler. As well, there are numerous other scientifically aligned books on personality disorders, such as Theodore Millon's (1995) classic *Disorders of Personality: DSM IV and Beyond*.

2 Ian Kershaw, 1998, October 1, Frankfurter Allgemeine Zeitung. Cited in Machtan (2001).

3 See Hitler's *Mein Kampf* (German edition, Munich, Eher, 1939). Cited in Schweitzer (1984).

4   American psychiatrist Paul McHugh has recently argued similarly about the euthanasia activist Jack Kevorkian (McHugh, 2006).

CHAPTER SEVEN

• ◦ • ◦ •

•••••

The Rise of the
Charismatic Leader

... urging Fidel Castro to abandon fields and castles
Leave it all and like a man
Come back to nothing special
Such as waiting rooms and ticket lines ...

Leonard Cohen
('Field Commander Cohen', 1974)

ombining Heinz Kohut's psychoanalytic insights with those derived from psychiatry and scientific psychology, and applying these to the biographies of prominent charismatic personalities such as Churchill, Freud and Hitler, ought to provide materials for an outline of the natural history of the charismatic personality. This sketch may then be used to diagnose and explain other charismatic individuals, although no doubt much fine-tuning will be needed to account for the many individual variations and situational differences encountered. As a loose guide, the natural history of the shaman, with its five stages comprising disturbed childhood, apprenticeship and wilderness training, quasi-mystical awakening, shamanic career and demise, can be used to fit the pieces together. Charismatic personalities have often been likened to shamans, although the shamanic role is somewhat culture-bound. To complete the picture, several other themes need to be added to take account of the differences between shamanic societies and more complex social organisations, specifically, the False Self (an entity that may not arise in shamanic societies), the special function of creativity, and what will be described as the Charismatic Predicament. As well, the story needs to begin before birth with the genetic factors that influence development. This chapter begins such a natural history.

Genetic Inheritance

It is now widely accepted that children are born with inherited psychological dispositions. Several personality disorders have been shown to be heritable to significant degrees (Thapar & McGuffin, 1993), and numerous traits, including religious interests, attitudes and values, and even political participation, may also be strongly influenced by genetic factors (Settle, Dawes, & Fowler, 2009; Waller, Kojetin, Bouchard, Lykken, & Tellegen, 1990). A number of the traits underlying the personality disorders, specifically anxiousness, callousness, cognitive distortion, compulsivity, identity problems, oppositionality, rejection, restricted expression, social avoidance, stimulus seeking and suspiciousness all have quite high heritabilities (Livesley, Jackson, & Schroeder, 1992).

Some inherited dispositions, while inclining rather than determining, nevertheless do tend to generate the kinds of environments in which personality disorders develop, thus becoming almost self-fulfilling prophecies. For example, an individual who is temperamentally suspicious tends to view others as suspicious also, and they in turn may come to treat him with suspicion, which makes him even more suspicious, leading ultimately to a climate of suspicion that fosters the development of a paranoid personality (Kiesler, 1996; Millon et al., 2004). In a healthy environment such an outcome would be avoided by the counter-balancing influences of the traits of others, but in a family that is stressed by conflict, abuse, grief, trauma, poverty or some other pathology, the more negative influences may prevail.

However, while this information serves as useful background and is suggestive of there being a biological substrate to personality disorders generally, when it comes to the genetics of narcissism and charisma, very little is known. The most that can be suggested is that certain traits such as optimism, creativity, high energy and high intelligence, and especially sociability (Korenberg, 2009), all of which are very likely to be strongly heritable, probably contribute to narcissistic and charismatic development.

Childhood

The childhoods of charismatic leaders vary greatly, as do the circumstances in which they are raised. Some are born into poverty and may emerge from childhood resentful and scarred, while others are born into comfort and security and may emerge with a fairly healthy socialisation. But more important than the external circumstances are their internal circumstances, their attitudes to themselves and to those they meet, and how they make sense of their experiences (their 'object relations' in psychoanalytic terms), and these are primarily determined by their earliest familial experiences.

Kohut has offered two scenarios to account for the development of the charismatic personality. In the first, involving compensatory grandiosity, the infant is inconsistently and unpredictably related to by its caregiver(s) in alternatively loving and indifferent ways. There may even be significant abuse or neglect or conflict in the family-of-origin, although these are usually not so extreme as to severely traumatise or impair the child. Rather, they incline him or her towards narcissistic thought processes and behaviours as a compensation for the familial deficits experienced, and as a defence against feelings of emptiness, unreality and unwantedness. Such compensatory grandiosity may also become suffused with rage, and this is also likely to be suppressed because to express rage in such an environment might be dangerous. Instead, the developing child crafts a functional persona and a heroic life script, which is ultimately aimed at gaining some substitute for parental love; that is, narcissistic supply in some form, such as adulation, applause and dominance.

The biographies of Churchill, Freud and Hitler conform closely to this scenario. These men suffered significant neglect in their childhoods, accompanied by baby worship — they were feted rather than raised — and a sense of privilege and elevation, if only as the child of very important parents in Churchill's case, or as a 'wunderkind' in Freud's case. This special treatment continued into adulthood, and although painful reality intruded often and early for both of them, the quality of the relationship each experienced with their parents in their minds (rather than in objective reality), was of a prolonged and unrealistically infantile yet 'heroic' quality.[1]

For Churchill this involved a rich fantasy life that he constructed to protect himself against hurtful reality, whereas Freud's fantasy was provided for him by his parents specifically to compensate for their shortcomings. Churchill dared not look at his parents' shortcomings, whereas Freud's parents' failings were painfully obvious in their one-room apartment. In both cases the result was a powerful mental script that drove these men to compensate for their parents' failings by powerfully striving for worldly success.

When little Winston was crying for attention he still knew that a great destiny lay before him, and that his mother 'shone like the Evening Star'. This was because he had created for himself the compensatory, grandiose dream that his mother truly loved him with a special quality of love, and that her unavailability was due to the very important work she had to do. Kohut has suggested that there may be a particular unconscious fantasy that underpins such a state: 'You (the carer) are perfect and I am part of you' (Kohut, 1966, p. 250). Similarly, when young Sigmund was alone because his parents were preoccupied with his siblings, he nevertheless still felt singular and elevated. Kohut suggests that the unconscious fantasy supporting this state may be: 'I am God' (or perhaps even, 'Mommy and I are One'; Silverman & Weinberger, 1968). The result in both these cases was the emergence of a young adult who needed to dominate others and to feel special, and who contrived to realise a grandiose destiny in order to recreate for himself the only relationships within which he could feel safe and comfortable. Such is the genesis of the 'man of destiny'.

Little Adolf's fantasy life was quite another matter, for a dangerous enemy lurked within his inner world. In a family already saturated with grief and loss he experienced trauma, fear and humiliation from his earliest years. Whatever grandiose fantasies he had of himself incorporated an acceptance of frequent death, a familiarity with random violence, an intimate circle of others whose dispositions oscillated between the poles of random violence and madness, and where his survival depended upon unceasing vigilance and cold calculation. It was to lead to a deep sense of grievance, a towering rage and finally to a bloody destiny.

The second of Kohut's scenarios involves fixated grandiosity. It results from a failure to mature emotionally past a particular stage of development, probably because of an overinvolved parent (or parents), whose 'baby worship' maintains and inflates the infant's primary narcissism beyond what is optimum for social development. In this scenario certain primitive relationship qualities, such as oneness, grandiosity and exhibitionism become fixated in the child's mind, so that he or she comes to expect the entire world to relate to him or her as this parent does. Such an adult learns how to behave normally, but beneath this behaviour there exists a strong need to feel elevated and an intense desire to be treated as special, a tendency to demand attention and a comfort with adulation. Hence he cultivates relationships that allow this. There is also an underlying anger, both at the overinvolved parent and at the world that consistently ignores his narcissistic claims. This anger is kept within bounds, and he employs whatever talents he has to charm, bribe, bully, manipulate or recruit others into giving him what he needs, but if all this fails his narcissistic rage may spill out.

If the developing child is sufficiently resourceful to survive in this way, (and many are not and become dysfunctional), then he or she enters young adulthood searching for some opportunity to recreate again the early parental dynamics in which they felt special, loved and secure. This scenario of fixated grandiosity may occur less frequently than the scenario of compensatory grandiosity, but Franklin Delano Roosevelt is an example of such a developmental trajectory, and will be discussed below.

Youth and Young Adulthood

This is perhaps the most interesting, and certainly the most variegated time, in the lives of charismatic personalities, akin to apprenticeship and time spent in the wilderness in shamanic mythology. The central task of this time is to adapt the narcissistic mentality to the demands and protocols of the world. Some individuals adapt fairly quickly, as Churchill did, achieving fame while still young.

Freud also found quick success but in a limited, academic way and without fame. Hitler was a complete failure until he entered the army, where he at last found a social structure within which he could excel. Obviously socioeconomic status has a lot to do with this. Churchill had all the opportunities and privileges of wealth and class at his disposal. Freud had fewer resources, although he came from an upwardly mobile family and was part of a disciplined and educated community. Hitler came from far less healthy origins, and the aspirant mentality in his family, such as it may once have been (his father had raised himself up), was overwhelmed by traumatic loss.

It may also be that the severity of the narcissistic mindset influences how quickly and in what manner one adapts to the world. Most of the charismatic prophets and gurus I studied in *Prophetic Charisma* seemed to need some kind of refuge such as the church or a supportive and unquestioning wife during this time, and none achieved fame while young (Oakes, 1997). Then again, these were individuals whose subsequent claims were extreme in the extreme, such as being God incarnate or God's messenger. Probably there is some interaction between the richness of the environment that one has access to, and the severity of the narcissistic mentality, that smoothes or complicates one's path.

What almost no charismatic personality does is settle comfortably into a humdrum life. Churchill was bursting to make his mark on the world, Freud was working hard to advance himself while destroying his compromising papers in order to throw his future biographers off the scent, and Hitler drifted through the bohemian subculture searching for openings for his unusual talents. For most charismatic personalities, youth and young adulthood seem to be a time of extended crises and energised questing. The developing narcissist stays open and 'seeking' for much longer than do most people, not settling on his life role until quite late. He may reinvent himself several times in this period as did Hitler, moving from aspiring artist to heroic soldier to revolutionary to politician. Or he may repeatedly apprentice himself to mentor figures as did Freud, searching for some magical key in each of the great men he emulated. With Churchill there is more of a sense of him expanding in all directions at once; the successful cavalry officer, the writer, and the politician all developed simultaneously in his thinking, as if he served three apprentice-ships in quick succession.

Again, depending on their access to resources, there may be stop-start development, or seamlessly consistent progress. Periods of confusion, loss of direction, searching, overcompensation and denial may alternate with times of highly energised work towards some goal, which may then be abandoned for another fallow period, or the energised pursuit of a different goal. While this also describes how many people pass through this time, the distinctive thing about charismatic personalities is that they seem to remain questing and open, in an almost adolescent sense, for much longer than do ordinary men and women. They do not crystallise their selves into conventional roles and personae early.

For Churchill, youth and young adulthood was a time of struggle to rise above the difficulties of his childhood. He discovered outstanding talents such as his powerful memory and a quality of fearlessness, and his love of language gave him a way of advancing in the world so he could realistically plan for achievement and success. He began to get a handle on the world. In less than 10 years he went from being the class dunce to becoming a modestly accomplished young man who, after repeated efforts, gained entry to the military academy where he continued to apply himself. Eventually, he graduated near the top of his class. All the while he polished his writing and dreamed of becoming a great hero. The psychological blocks that he had experienced as a child such as anxiety, inability to succeed and a painful shyness around females he was now overcoming by dint of sheer energy and will, although the last of these blocks took a few more years to overcome. When he finally saw military action the pent-up energy fuelled by years of frustration was set loose as he seized his chance to become the hero he had dreamed all his life of becoming. He distinguished himself in combat, then sang his own praises in newspaper articles and books he wrote from the front. His escape from the Boers consolidated this heroic image, both in his own mind and in the public's appreciation of him. But at an unconscious level what he was really engaged upon was a do-or-die effort at gaining his parents' love.

Churchill was living out the personal myth of calling he had constructed as a child, as the hero called by destiny to achieve great things. Such success probably also reinforced certain infantile attitudes he had retained from childhood, specifically a kind of magical thinking, as if he was blessed and needed only to 'think and make it true'. This, in turn, gave him boundless confidence. Where once he had anxiously hoped for success in some distant future, now he was enjoying it daily — anything might be possible. He cut some corners however, earning disapproval from senior military commanders whose protocols he flouted, and his ceaseless self-promotion was considered vulgar by many of his peers. But he placed himself above such considerations. While being too inhibited to behave antisocially (and having no real need to), nevertheless there was an amoral opportunism to his nature that was not above raising himself at the expense of others, as the controversy over his escape from the Boers showed.

For Freud, youth was much more about living up to his parents' expectations in a conventional sense. He lived out the myth of calling inculcated by his parents during his childhood, and he developed his prodigious intellect. He studied, he excelled and he also cut corners professionally, and even ethically. He had tremendous confidence in himself and took great pains to study widely and to seek mentors and guides, yet he was also depressed and confused frequently, and he made several major mistakes. If his thinking was not very 'magical' at this time, nevertheless he pursued the magical and he invested his mentors with almost magical properties, as if each held a key to some great secret.

Through his scientific work, his experimentation with cocaine and his friendships with mentors, Freud began to break out of his emotional blocks (Breger, 2000, p. 69), and to move from fulfilling his parents' hopes for him to developing his own heroic agenda. Freud's form of self-promotion involved publishing scientific papers, some of which had lasting significance, and even though the topics he researched had little to do with his later interests (he was for some time engaged in a study of the testicles of eels), the skills he gained and the professional connections he made in this time proved to be of lasting value.

For both Churchill and Freud, young adulthood was a time of conscious preparation for the destiny that each believed lay before him but could not see clearly yet. Both had tremendous energy, worked extremely hard and were highly oppositional, flouting established protocols when it suited them. For each there was a delayed entry into adult life, especially into marriage, suggesting that the transition from child to man did not flow easily, probably because each needed time to figure out how to adapt their narcissistic mindsets to an indifferent social world. Each was dogged by emotional difficulties, mostly depression, and sought support in drugs. Each shaped his early manhood into a great creative project that he expected would propel him above the mass of lesser humanity.

In contrast, Hitler was quite lost for most of his youth and young adulthood, and this was probably a direct result of his much more troubled childhood. He lazed around, largely because his way forward was problematic; he received no obvious pathway forward from his parents or his class or his community, and he failed in his effort to become an artist. His homosexuality would have also been a compounding factor, something to keep secret, an added complication. He spent some time on the run, living from hand to mouth, mixing in a bohemian subculture that accepted him simply because it accepted everyone. He dabbled in fantastic schemes such as rebuilding Berlin and composing an opera, and he was not above spiriting away some of his sister's inheritance. He finally settled in the army (it also accepted everyone), where he won two Iron Crosses. These honours set him up to work as a government agent and made possible his subsequent entry into politics, which may not have been his aspiration at the time. He was socially awkward and lacking in leadership qualities when young. However, his work as an undercover agent led to useful contacts, and to access to the lower levels of politics. He finally found his way forward when he discovered his rhetorical ability, but even then he still made a mistake with the failed Beer Hall Putsch, which led to a spell in prison. Despite his awakening in Pasewalk Hospital in 1918, only in his mid-thirties did his destiny as a politician finally become clear to him. Not for nothing did he title his book *Mein Kampf* [My Struggle].

Other themes active in the young adulthoods of all three men include a radical autonomy from society's conventions. Churchill hacked his own path through the respectable protocols of his class and time, and Freud unashamedly fudged his research results to pursue fame. Hitler lived a very unconventional life in the homosexual subculture, the nature of which has only recently become known. Each chose to live according to their own feelings and intuitions (or perhaps they really felt they had no choice), following their own myth of calling rather than deferring to reference groups and cultural norms. They cooperated with the conventions of their day only insofar as these were useful; they did not feel bound by them. In Churchill's crossing the floor of parliament to join Lloyd George's Liberals, in Freud's indifference to his Jewishness and in Hitler's homosexuality there lies a kind of alienation, a sense of not belonging to any group, or at least of escaping the confines of being defined by or bound to one's sociocultural origins.

For these leaders young adulthood was also a time of chronic conflict with authority, almost a sine qua non of charisma. Churchill seemed to enjoy bucking the military and parliamentary systems, whereas Freud, although not personally aggressive, nevertheless was happy to flout established scientific protocols, although as a vulnerable young Jewish professional he avoided direct conflict with his superiors. Hitler's young adulthood was lived in constant tension with the conventions of his time.

The splitting of the personality (Kohut, 1977; Oakes, 1997) that had begun in childhood with the construction of a false self, reached fruition in young adulthood as each glimpsed the special destiny that lay before him, and transformed and streamlined himself into a one-dimensional persona that embodied that destiny. Consistent, ruthless, energised pursuit of the newly revealed image of himself, and the special destiny that each gradually perceived, happened at the height of this stage of development and led into the great creative work that followed. There was massive variation in how each did this, but each did it as if he were following some deep agenda known only to himself. The false self is the lead actor in the creation of such a myth of calling.

Lastly, this is also the time of acquisition of the career skills, qualifications and symbols[2] that become so important in the later careers of charismatic personalities. Expertise in the communication arts is vital; Freud and Churchill were highly skilled writers and presenters of ideas, and Hitler mastered rhetoric. Charm and what has been called 'the giving orientation', an attitude of giving to others in order to get their love and support (Oakes, 1997, pp. 87–88), is also absorbed at this time, probably consciously and deliberately as a tool for self-promotion.

The leadership skills of Churchill, Freud and Hitler, especially in their younger years, were incomplete. Churchill's leadership skills were mostly just force of character; his threat to 'break' anyone who opposed him when he assumed command of his army unit in World War I was hardly subtle. Freud's leadership skills never really emerged until his great mission fired him, after which he became slyly manipulative and devious. Hitler's leadership skills did not appear while in the army.

It is likely that the narcissistic mindset may delay the development of leadership skills for many charismatic personalities. There are some who set out quite early to master leadership skills, individuals who get started early in a career in sales or the media, for example, but the three described herein seem to have had to learn as they went throughout life. This shows that charismatic personalities may not be natural leaders, paradoxical as this seems. They naturally live through others, but how to control and influence these others may be something that they learn slowly, on the job as it were. Most do not start out as deft, professional leaders; rather, they are inspired amateurs who opportunistically become leaders and then have to rise to the occasion.

Qualifications of one sort or another are also crucial, although they may not be qualifications in the usual sense. For Churchill, his fame as a soldier and writer set him up for electoral success, and Hitler's two Iron Crosses opened many doors. Imagery and symbolism are also important. Hitler experimenting with Tyrolean hats and a riding whip, and Freud's outright purchase of 'Professor' status, rounded out cultural touches that each needed on their way to their niches as 'Hitler, man of the people', and 'Herr Professor Freud'. Churchill's jaunty cigars were used to signal someone much more exciting and colourful than a shallow toff. What seems to happen is that successful charismatic leaders, along the way, manage to achieve one or more relative qualifications that they are able to exploit, more in a symbolic sense than as any literal credential.

For charismatic personalities, qualifications generally are more problematic than is usually recognised. Buckminster Fuller did not have the right credentials for the work he did, and Erik Erikson was basically an engineer rather than a social scientist. Harry Stack Sullivan, the 'father' of American psychiatry, had no real qualifications at all, only a bogus certificate from a diploma mill.

The life of Mohandas Gandhi may provide an even better example of this stage (Gandhi, 1927). Gandhi actually subtitled his autobiography, *The Story of my Experiments with Truth*, alluding to the seeking, questing nature of his development, and he spoke of living 'an experimental life'. For much of his life he was preoccupied with 'experiments' to do with everything from diet and hygiene to birthing and social relations, and his later actions with teenage female followers (in the cause of spiritual self-control over carnal desire, he claimed), should be seen in this light. Such experimentation was at its most intense during his adolescence and young adulthood.

Gandhi was, quite typically for a Hindu in those years, married at age 13. Soon after, he explored a variety of taboos including meat-eating, smoking, theft and visiting a brothel. This led to a painful life-changing experience that crystallised his values (see chapter 13), but he continued his experimentation. At 18 he went to England to study as a lawyer and there he conducted more experiments. In addition to studying law he enrolled in Latin and French classes and studied other subjects that were unnecessary for his law degree. He joined a vegetarian society and went on a jag of dietary experimentation. He read voraciously and undertook 'the all too impossible task of becoming an English gentleman', taking dancing lessons, attempting to learn the violin and taking lessons in elocution. The upshot was his discovery that, 'I was pursuing a false idea'. Yet he continued with his 'determination to make a complete change' (Gandhi, 1927, pp. 62–63).

Gandhi also experimented to see how simple he could make his life and how frugally he could live, shifting premises several times after negotiating more convenient and economical living arrangements. He travelled widely around London by walking because he was too poor for frequent public transport. He drew great satisfaction from such experiments, and the money saved enabled him to visit Paris to see the Great Exhibition and the Eiffel Tower in 1890, again mostly by foot. He investigated Theosophy, meeting Madame Blavatsky and Annie Besant. He read the Bible and came to a renewed appreciation of his own faith through this. He also worked on his personal limitations and attempted to overcome his shyness, undertaking public speaking activities and reframing his shyness as 'my shield and buckler', and a support in 'my discernment of truth' (Gandhi, 1927, p. 72).

Gandhi has written quite candidly of the veneer of deceptiveness in his persona that was necessary for such experimentation, giving chapters of his autobiography titles such as 'Playing the Husband' and 'Playing the English Gent'. This suggests that he was developing something of the street-smart mentality. He has also spoken of following his 'inner voice' through much of this (Gandhi, 1927, p. 57). As his biographer Louis Fischer has said, 'To the end of his days Gandhi attempted to master and remake himself ... [in a] remarkable lifelong task of changing his mind' (Fischer, 1997, pp. 39–40).

In South Africa several years later, his experiments bore fruit. Fischer has written that Gandhi remade himself there, undergoing a 'second birth', discovering his talent as an organiser and also an aptitude for nursing, acting as midwife in the births of his own children. By age 35 he had become successful and influential in his own right, having lived and worked in three continents. At 37 he took a vow of celibacy. By then he had transformed his circumstances and his personality, and his 'awesome Himalayan self-assurance' (Fischer, 1997, p. 101) began to appear, shining like a beacon to colonised peoples everywhere.

In sum, the period of adolescence and young adulthood is a time of great experimentation for charismatic personalities. Most make an effort to reconfigure their personalities, splitting off parts here, cultivating other aspects there and adopting images and roles. They are forced to do this because their narcissistic mindsets are so at odds with the indifferent world; they must find ways to recruit others to provide the narcissistic supply they are so dependent upon. This process, seemingly so driven, strange and creative, may in fact be merely an amplification of quite normal processes. Most personalities change over time and, contrary to earlier thinking, the time of greatest change for 'normal' personalities occurs between ages 20 to 40, when we become more stable, conscientious, agreeable, open to experience and — especially — dominant (Roberts & Mroczek, 2008).

Creative Awakening

Creative awakening is not so much a separate developmental stage, though it can be. Rather, it is more of an enduring theme that can have a profound zenith, usually sooner rather than later, and which corresponds to the quasi-mystical awakening of the shaman. The process may include an actual mystical experience for those charismatic personalities so disposed, or it may constitute a frenzied phase of creative activity. More commonly, however, it may include one or more great creative projects carried on over a lifespan, albeit usually with a time of maximum creative effort that signals something of a turning point in life, and which initiates significant personality and behaviour change. There may even arise a 'creative illness' such as that which Henri Ellenberger has described in the case of Freud (Ellenberger, 1970), at least for those 'sick-souled' individuals of a basically pessimistic turn of mind whom William James studied (James, 1961). For those charismatic personalities who are so inclined as to have full-blown mystical experiences, these nevertheless occur within a life context (they do not come out of nowhere), that includes a preceding, intense and ongoing creative effort, and which subsequently leads to self-transformation.

The claim of mystical or quasi-mystical experiences needs to be viewed in the light of recent social science scholarship that sees mystical awakening as a creative problem-solving strategy of a high order (Batson & Ventis, 1982). Those who argue for a special metaphysical or ontological status of the mystical experience face several objections. First, many deeply religious persons live lives of devout spiritual involvement without ever having a mystical experience, hence such experiences are not necessarily a part of religiosity. Second, the experiences themselves produce such divergent and conflicting reports as to be irreconcilable, and to strongly indicate that they are primarily psychological rather than metaphysical in nature. Third, there are other experiences that occasionally lead to much more interesting and meaningful reports, specifically some psychedelic and near-death experiences. Hence, the mystical experience need not be thought of as any kind of 'ultimate' encounter with truth, although that is how it is often experienced.

The argument advanced here is that through an intense and continuing creative effort, mystical or quasi-mystical experiences may or may not occur, depending upon the disposition of the individual involved. Regardless of this, great creative projects leading to extraordinary achievement and self-transformation may be realised, with or without such experiences, although they are quite common for such persons. But it is the ongoing creativity that is crucial here, not any single manifestation of it.

Churchill's great creative project entailed emulating his father's political career. He spent endless hours developing his rhetorical ability, rehearsing over and over, this way and that, how to construct and deliver phrases for best effect (and in the process overcoming his speech impediment). He became a brilliant orator, delivering from memory whole speeches in the House of Commons, and usually dazzling his opponents and besting his critics with his expertise. His caustic wit is legendary. He developed an uncanny ability to read the hearts and minds of his audience, presenting his material in ways that appealed to their needs and hopes, or that played upon their fears.

Equally important as his speechifying was Churchill's problem focus. He sought out opportunities to present himself as solving great problems, ever convinced that he was the best man for the job. He saw the opportunities presented by Lloyd George's government and crossed the floor to help develop the British welfare state. He volunteered for the most active fronts as a soldier. Throughout his career he unerringly homed in on the most difficult of problems, and when World War II broke out he sought and received the leadership of the nation.

Freud's early creative projects were restricted to research and study, but it is clear from his experimentation with cocaine, from his attachments to his mentors, and from his work with Breuer, that he too felt compelled to be constantly seeking and exploring; his early unpublished book on scientific psychology is one example of this. It all led up to his supreme creative act, his self-analysis under the mentorship of Wilhelm Fliess, including the 'creative illness' that many mystics, artists and shamans undergo, and from which he emerged with a new vision of the world (Ellenberger, 1970).

With this new vision Freud founded his Vienna Psychoanalytic Institute and became something of an evangelical prophet, publishing his latest revelations in his house journal, recruiting disciples and manipulating them this way and that, dealing with dissenters and defectors, and organising and managing what became a large network of followers. While doing all this, however, Freud found time to revise and expand his teachings and to continue to develop them into new areas, virtually inventing such fields as psychobiography, psychohistory and critical social theory with novel essays such as *Leonard da Vinci: A Memory from his Childhood, Moses and Monotheism, Civilisation and its Discontents* and *The Future of an Illusion*. The fact that he got much of this wrong need not be held against him, for it was the general thrust of his thinking that was most important. Besides, others followed him and produced more enduring works. In all it was a sustained outpouring of creative thought such as few thinkers before and none since have equalled.

Hitler was also highly creative. He developed Nazi ideology and pioneered the use of the media for propaganda. He was brilliantly successful on a grand scale in both domestic and international politics, and masterful in his grasp of military strategy, at least in his early campaigns, (although Sebastian Haffner [1979] has shown that these successes were all against weak opponents). This was an enormously diverse and impressive achievement.

One of the most intriguing things about charismatic personalities is their ability to recreate themselves, not merely once but perhaps several times if need be, throughout their lives, as they reinvent themselves to exploit the opportunities their creativity opens up. In Freud's case he changed his personality dramatically after his self-analysis and the publication of *The Interpretation of Dreams*,

(Ellenberger, 1970). In Churchill's case, once he saw clearly the threat posed by the Nazis, he also changed his behaviours as he took up his mission to protect the European democracies. Hitler prided himself on his refusal to change, but the man who became Chancellor of Germany in 1933 was quite different from the 30-year-old former soldier of 1919. At that time he had been confused and demoralised, jobless and without prospects, lacking in confidence and anxious about his future, but in the next 14 years he transformed himself, effectively becoming all at once a full-time actor, an intimidating schemer, a skilled political opportunist, a sublime manipulator of crowds and ultimately, a mass murderer without conscience.

For each of these men personality change was closely related to their creative efforts. For Freud, personality change followed immediately after his creative achievements, while for Churchill it followed his creative insight into the enormity of the danger the western nations faced. For Hitler, personality change was part of a drive to gain power and manipulate the crowd. In each instance it was as if they glimpsed some great chance, held nothing of themselves back from it, and did whatever was necessary to achieve what they had set their sights upon, including transforming their personalities.

Of course, it can be argued that none of these leaders really changed their basic natures, and this is probably true. Rather, they adopted roles and presented public images and cleverly managed the impressions they made on others. Yet so 'natural' did these changes become over time, so totally did the false self overtake the real self, and so unreservedly were these people committed to the changes each had made, that for practical purposes there is little to distinguish between the reality and the image. As Leonard Bernstein said of Gustav Mahler, he was not a great man; rather, he was a great actor portraying a great man, but it comes to the same thing.

The corollary of this is that when one has invested so much of oneself — one's very identity — into one's creative efforts, then if these efforts are subsequently invalidated one feels as if one's self has also been invalidated. Hence, there can be no going back once one has made such a commitment. One defends one's vision to the death, as if one's very self is sustained by it, because emotionally it is. In some corner of his mind Freud probably guessed that classical psychoanalysis did not work very well as a therapy, though he had his successes with it, mostly due to his charismatic presence. To his credit he modified his hopes for it in his 1937 essay *Analysis Terminable and Interminable*. But he just could not bring himself to face the fact that his central concepts were wrong, and he kept promoting his illusory constructs until the day he died. Churchill clung on to power long after he had been rendered ineffective by stroke, and he kept his parliamentary seat well into his infirmity. Similarly, Hitler could not face defeat, and he kept issuing orders to nonexistent armies and talking up his imaginary secret weapons until the last days.

Effectively, each of these men became their great creative project, living and dying by it, forsaking other attachments for it and subordinating their beings to it in pursuit of the ultimate success that each unconsciously hoped would truly heal them. Each identified so totally with their creative project — Churchill with his historic destiny, Freud with his system of psychoanalysis, and Hitler with his Nazi vision — that they neglected or suppressed or massively split off their real self, or at least those parts of it that did not fit into their creative project. The seeds for such splitting of the personality had been sown in childhood, had reached fruition in young adulthood and then grew to consume them in their mature years, rendering each a one-dimensional persona in which the real self had been subordinated to their function. In *Prophetic Charisma* I reported how one guru told me, 'I've become the message now …', and another said, 'I am just my function' (Oakes, 1997, p. 13). Great charismatic personalities seem to lose themselves in their missions. Their all-or-nothing, life-and-death, 'crash-through-or-crash' commitments demand that nothing of their real selves is allowed to impede their great work. Even the lesser ones do this to some degree. Kohut has referred to this as abandoning 'the core of the self' (Kohut, 1971, p. 117). To what extent the real self remains reclaimable in later life is considered in chapter 8.

Mission

Paralleling the shamanic career, this period follows from a time of intense creativity that has resulted in significant personality, or at least behavioural, change. The creative effort will have opened new opportunities and possibilities, and through these the charismatic personality will also have attracted a small following. Maturity will probably see him continue his creative and transformative process. But 'mission' really signals the charismatic personality at the height of his powers, leading a movement dedicated to fulfilling his vision of transformation or world salvation or, for those many lesser charismatic personalities, anything from church reform to administrative change, or from community development to winning the cup.

Despite appearances, this is not really a time when he is completely in control, or only just barely, and his mission may not bring out the best in him. He has become successful, but now he must exploit his success while satisfying the demands of his followers, playing their game as it were. If he feels stressed or exhausted he nevertheless must still perform well while under close scrutiny.

Most notable in this time is the astonishingly high energy that the charismatic personality — by now the charismatic leader — is able to draw on in a crisis. He is able to apply himself to intense and sustained work to achieve desired ends, and he accompanies this with resourcefulness and creativity that inspires those around him. There is a sense of rising to face an ultimate challenge. Churchill's youth and middle years were very active, but not as intensely active as his war years. Similarly, Freud spent his younger years in intense creative work, but he continued such creativity unabated while also founding and leading his movement and while seeing patients 10 to 12 hours per day, six or even seven days each week for the remainder of his life. Much has been made of Hitler's youthful — and even later — laziness, but he too was capable of sustained bursts of activity right up until the end, though most especially during the period 1930–1933 when he was travelling the country in his aeroplane drumming up support (Waite, 1977).

This is a time of great confidence and heroic goals; when the charismatic leader is constantly talking himself and his followers up. He exudes a certainty as to his ultimate success, his exceptional personal qualities and the rightness of his cause. This is closely related to his sense of entitlement, for he truly believes that he has the equivalent of a divine calling or a historical mission, a duty to lead, advise and teach others (which is also a need to be raised above them), and he expects compliance and gratitude.

However, it is also a very demanding time for it requires the mastery of the skills of leadership, something the charismatic personality may not have bothered with up till now. Many charismatic leaders are defeated by the number and complexity of the management skills they have to learn in this mission stage. Politics and religion are littered with failed demagogues and prophets who inspired followings but never really developed the skills necessary to manage them successfully. Those who do so have usually cultivated communication skills and experienced some minor leadership roles before. The commonest mistake is delegating to unreliable subordinates. The leader needs a competent executive, but this is not created overnight. Churchill inherited a state structure with which he was already familiar, and Hitler let his lieutenants create their own structures, leading to a fragmentation of the state and a proliferation of bureaucracies and even semiprivate armies. Freud somehow found time to study and master interpersonal manipulation in order to hold his cult together, although there were many defections and inner tensions.

By this time the leader will probably also have acquired other important attributes, such as identifying themselves strongly with a particular image that others strongly relate to, typically the man-of-the-people or the aristocratic altruist or the virtuous scholar or whatever. They will also have accumulated a body of specific knowledge that qualifies them as an expert, at least to their followers in an hour of crisis (Oakes, 1997). Charismatic leaders do not spring from nowhere, and while some of their 'professional development' may have been accidental insofar as they may have studied communication or management for career purposes, such skills are crucial for later success.

If astute, by this time the leader will have engaged the support of key others to ease the burden of leadership. The loyal lieutenant is one such figure. The charismatic personality usually leaves a lot of chaos, confusion and hurt in their wake. It takes a committed administrator-cum-housekeeper-cum-counsellor to smooth the ruffled feathers and soothe the bruised egos of followers who, without such placating, might angrily defect and cause incalculable damage because of resentment at the leader's shabby treatment of them. Charismatic personalities generate as much hate as they inspire love, and dramatic betrayals are by no means uncommon. Such a loyal lieutenant may even come to exercise a modicum of influence over the leader; by consistently demonstrating their fidelity, they achieve a degree of trust and intimacy that no one else attains, although really, no one has any strong influence on the narcissistic personality because he is too self-contained and self-sufficient. But the lieutenant may be accepted as an integral part of the refining or winnowing process for developing the leader's ideas. One of Churchill's adjutants has said that he (Churchill) had about 20 brilliant ideas each day, and that his (the adjutant's) task was to figure out which one or two might be practicable.

Another key recruit is the undemanding spouse who is satisfied with a lesser role, and who accepts the firm limits that the leader places upon him or her in return for security. Clementine Churchill and Martha Freud accepted their husband's preoccupations with their charismatic vocation as a trade-off for their own family security, whereas Eva Braun lived in hope. The charismatic's wife is not quite a doormat, but nearly. She is able to sharply delineate between her own sphere of interest and his, and to stay well out of his, while accepting only token involvement from him in hers. Although she may be disappointed in some aspects of the marriage, she stays because he is the most fascinating and stimulating person she has ever met. She is certainly not merely docile; a more complex relationship is involved. There may be love (of some kind) on both sides of the relationship which, if for some reason it suddenly ends, may lead to the wife remaining single for the rest of her life because she knows that no one else could ever be as exciting as her departed spouse.

There are other intriguing roles within the successful charismatic movement, and not all of them are functional. There is the special follower upon whom the leader pins a lot of his hopes, and who often proves unworthy. There is the honorary Judas who can be blamed for mistakes when things go wrong, and who appears to accept such a role, while in fact having much more complex motives. There is the cheer squad, who can be counted on to shout down opposition to the leader. There are highly placed outsiders whose support enhances the leader's credibility, and there is the great external 'enemy' who exists 'out there', and whose alleged threat galvanises support around the leader and enforces a sense of crisis and dependency that the leader exploits.

The burden of leadership is further softened by the nature of the charismatic relationship, which is ultimately not about outcomes but is rooted in love and hope, and thus is somewhat uncritical. The followers do not follow just for worldly gain; they hope for personal transformation rather than merely a profitable transaction. They follow the leader because they have come to love him for his strength and wisdom, and for his flattering vision that ennobles their prosaic lives. For by now the charismatic personality has become an adroit social manipulator with an uncanny ability to read the hopes, needs and fears of others, and to exploit these with rhetoric, symbols and images — the language of the crowd.

In his relations with his followers the leader comes to prize loyalty above all else, and he cultivates an intensely personal commitment in the followers, a commitment that overlooks mistakes, in faith that the leader truly loves his followers. As the good Germans protested when they learned of the Nazi euthanasia activities, 'If only the Fuhrer knew about this'. An important part of this is the 'charismatic claim' made by the leader to the total allegiance of the followers, and to be the sole source of their ultimate good (Oakes, 1997, pp. 115–118). The sheer boldness of the claim is part of its appeal, for it is invariably couched in uncompromising terms. The leader asserts himself as the only vehicle for the realisation of the followers' salvation, or the advancement of their hopes, and proclaims his unstinting and unswerving devotion to them. As Jesus said, 'I am the way, the truth and the life', or as many a charismatic leader since has proposed, 'Everything I do, I do for the people'. Not all charismatic leaders get to advance this claim; in some milieus such a claim might be unsayable because of the followers' worldliness, but probably each believes it deep down. In a time of crisis such a totalistic claim may be necessary to stand out from rivals and the crowd, and at such a time it inspires faith, hope and love in the followers, as if the leader is calling them to also make a total and uncompromising commitment as he has done.

There is great variation in the forms of charismatic leadership, because this time of mission is not solely within the leader's control, but is largely determined by external events. Some of the more influential factors include the number and natures of the followers attracted, whether the movement develops swiftly or slowly, conflicts with external authorities, power struggles within the movement, the influence of dissenters and apostates, and any sudden or unexpected opportunities that may arise to advance the cause (say, a large inheritance coming their way). In a successful charismatic movement the leader's resourcefulness will probably be tested at some time by all of these events occurring. The most frequently encountered problem of charismatic religious movements is rapid growth leading to the promotion of unworthy second-level leaders. Some charismatic movements are defeated early by conflict with outsiders, typically many African charismatic movements in the postcolonial period, whereas others such as the Nazis are set up specifically for such conflict. A power struggle within his movement destroyed Indonesian leader Sukarno, and J.H. Noyes was derailed by dissenters within his Oneida Community. Other charismatic leaders are seduced by wealth in the midst of their programs, typically American televangelists. Each of these eventualities pressures the charismatic leader to adapt and behave in ways that he or she may not have prepared for, to react rather than to proactively plan a campaign, and to make up policy on the run, rather than to think things through in advance. There are dangers inherent in each of these options.

So demanding can the leadership role be that the charismatic leader risks becoming preoccupied with issues of power and image, sacrificing even his message for success and to maintain his following. Success may become as big a problem for the charismatic personality as failure might be. For it is hard to forego that satisfying applause just for one's principles, no matter how much one has invested oneself in them. There are few sadder individuals than a corrupted and forsaken ex-charismatic leader. There is a unique quality of hollowness, haunted confusion and desperation to such figures once they have been abandoned by all around them. It is as if they sold their very souls for power, only to be betrayed by something they must never comprehend — their own limitations.

Another risk is that the leader may lose touch with his followers by becoming too involved in large-scale concerns. The leader's power ultimately comes from the people, and if he neglects them he risks losing everything. It is not commonly recognised that leaders of charismatic religious movements do not totally immerse themselves in esoteric mumbo-jumbo, but often work hard to make the lives of their followers more fulfilling and successful. It is this practical side to charisma that is often overlooked. This frequent demonstration of the common touch is shown by Freud's involvement in the lives of his analytic trainees, Churchill's visits to the sites of bombed Londoners and Hitler's tours among Germany's youth. In sum, the followers' allegiance is not totally uncritical, and may stray if their lives do not improve in practical ways through following their leader.

Endnotes

1   This chapter has been written in naturalistic terms, but in pure psycho-analytic theory it is not so much the objective reality of family relationships that is important as it is the internalised 'self-objects' within the developing child's mind, (and which correspond to external family members). Because there is usually a great deal of overlap between the real persons and the child's experience of them — that is, the external mother really is like the image of the mother in the child's mind — it is permissible (I believe) to conflate the two in an account such as this. Nevertheless, it raises an interesting question about whether a charismatic personality could arise in someone whose experience of life was very different from the reality of the external world. Might he construe a consistently indifferent parent as a loving one in an act of compensatory grandiosity and emerge as a charismatic personality? I believe that this could only happen if the person was psychotic or semipsychotic; being so detached from consensual social reality is part of the definition of psychosis. Probably, because charismatic development is so problematic it could not even begin if there was psychosis present. Then again, a supremely gifted person such as Hitler might just be capable of such psy-

chological prestidigitation. However, even so they would surely emerge into adulthood with traces of the original psychotic mindset. In this regard it is instructive that Kohut theorised that Hitler suffered from a 'healed-over psychosis' (Kohut, 1971, p. 256).

2   Charismatic symbolism was originally characterised as 'It' by Elinor Glyn, the British romance writer who coined the term in 1927. Actress Clara Bow was the first 'It' girl; she had 'It' in abundance. Glyn wrote that, 'In the animal world, 'It' demonstrates itself in tigers and cats, both animals being fascinating and mysterious, and quite unbiddable' (Roach, 2007, p. 4).

———————————

• • • • •

CHAPTER EIGHT

• • • • •

Decline or Fall?

Virtually everyone leaves Utopia after a time. The quick and hearty do not necessarily defect early, nor is it always the witless who linger on. One leaves when he has gained what he came for, when his commitment is exhausted, when it is no longer necessary to sort through the breviary of questions that concern his freedom. There are only four seasons in Utopia and they are not repeated. Therefore the spring is more vivid and the winter meaner. What is difficult to remember is the first day of fall.

Tom Patton (1980)

The completion — or otherwise — of the leader's mission, depends on so many factors as to make it too unwieldy to summarise here; success and failure are only two considerations among many. Fortunately, as the focus of this book is the psychology of the charismatic personality, there is no need to. If the mission fails then there is likely to be little or nothing left to explain. But if the mission succeeds then a range of novel possibilities may arise, and most entail risks. For at base, the leader's mission is a means to an end, a program to reinstate the emotional life that he enjoyed in his formative years. But everything comes at a cost. Success may have cost the leader dearly; there may exist little or nothing of his former life left to reinstate. If so, then he may be tempted to remain with his mission, pushing it ever further in pursuit of some elusive redemption. He may become lost in it, believing totally in the role he has created, and dependent upon the narcissistic supply it brings, all the while becoming increasingly frustrated and empty, although never knowing why.

The False Self

The false self is not really a stage of development as much as an orientation to the world that becomes habitual. It is an enduring persona that forms during adolescence or young adulthood and reshapes itself anew in different situations, most especially in maturity when it becomes deeply rooted in the personality (Masterson, 1988). It is now recognised that personality changes greatly during the life-span, and it changes most during our third and fourth decades (Roberts & Mroczek, 2008). Hence, it is not really so unusual for the developing narcissist to attempt to remake himself in these years. That he does so deliberately and in accordance with an image he has of himself as a leader is perhaps somewhat uncommon, but probably no more so than how many people cultivate an image of themselves that is intended to assist them to get ahead in life. The main indicators of the false self in the lives of Churchill, Freud and Hitler include the following traits and behaviours.

First, there was a failure of self-understanding. From his earliest years Churchill believed he was the hero of his dreams, when in reality he was a sad and lonely child. This became a blind spot in his self-image; he could not acknowledge how hurt he felt at his parents' neglect, and he mentally split this part of himself off so successfully that he was able to live most of the rest of his life as the persona that he invented for himself. However, there were times when his splitting failed and he encountered his 'black dog' of depression, when he was unable to live his heroic script, typically his exile years between World War I and II. Then, although he still had everything going for him as a writer, husband, father, parliamentarian, aristocrat and famous raconteur, he was denied the opportunity to live out his heroic role and he regressed to that aching, neglected child.

Freud similarly sought fame as a great scientist, but in the years before he became famous he was driven by his parents' script for him as a wunderkind, a script devised to compensate for their own shortcomings, and whenever this script was insufficient to sustain him, he too lapsed into depression. When Freud underwent his self-analysis he came up with his fanciful sexual aetiology for his neurosis, a narrative that ennobled him through the mastery of rational self-control over base animal nature, while it neatly avoided consideration of the real causes of his anguish.

Hitler probably had no insight whatsoever into himself, far less than Freud or Churchill even at their most misguided. More than either of them he lived in a world of his own projections that was wrong about most things. The false picture he created of his childhood in *Mein Kampf* was concocted for propaganda purposes, but he may have come to believe in it (Erikson, 1942).

The second indicator of the false self is grandiosity. Each of these three figures had an inflated sense of self-importance, saw themselves as special in some sense, and expected to be recognised as superior. Their grandiosity was so transcendent that it enabled them to override all manner of cultural and personal taboos, to risk and to succeed where others would shrink back, but ultimately it did not serve them well. The tragedy of Freud's marriage was that he was never able to replace the lost sexual satisfactions of his early years with the soulmate function in his later years; his need to feel strong and grandiose prevented such a close relationship from developing. Churchill's marriage also seems to have never been really close, and though this may have been typical of marriages of that time and class, nevertheless it clearly suited his grandiosity to have a wife as an adjunct to his role as a leader, rather than as a genuine intimate. Hitler's megalomania has become synonymous with his name.

Third, and arising from their grandiosity, was their lack of close, equal and intimate friendships that might serve some correcting or humanising function. Close scrutiny of the friendship patterns of each of these men reveals that their 'friends' served psychological functions as admirers, followers, patrons or supporters, but were not really taken into confidence openly and honestly. They were sources of advancement or narcissistic supply, rather than confidantes and equals. As Kohut has advised, the charismatic personality uses others as parts of himself, to provide balance, comfort, support and a flattering reflected self-image, rather than to experience any kind of I–Thou communion.

Fourth, related to both their grandiosity and their lack of close equal friendships, is a cluster of neurotic behaviours around issues of power, including an inability to accept criticism or to tolerate dissent, and a need to be right, or at least to feel in control. The false self cannot admit fault or vulnerability or any kind of shortcoming; some extremely narcissistic therapy clients are unable even to ask for directions or to admit to any lack of knowledge on their parts, even if it is about something that they would not normally be expected to know such as a scientific term, so intense is their need to appear strong, able and self-sufficient. It is as if they have 'one skin missing', so hypersensitive are they. Again, this point virtually needs no illustration for the three figures presented herein, each of whom was extremely thin-skinned, had a legendary need to be right and to define reality for those around them, and to exercise control. Each had the proverbial monstrous ego.

But such dominance can be a trap, for even if others acquiesce for the sake of peace, they do so reluctantly, watchfully and, in difficult times, resentfully. Such a stance also has to be constantly defended. Probably nobody would adopt such a position unless at bottom they felt very insecure (despite whatever disguises they had in place). It also suggests a fairly crude understanding of human behaviour; a belief that people can be intimidated or conned into loyalty. Only someone who was needy and who possessed both exceptional ability and extraordinary short-sightedness would try to make such a stance work for them.

Fifth, such individuals need an enemy. This may be a specific group such as the Jews were for Hitler, or a threatening power such as the Axis countries were for Churchill, or the forces of ignorance and superstition as with Freud. Or it may be an institution such as the Church, a threatening socioeconomic situation as with cargo cults, or even some oppressive psychospiritual doctrine such as emotional repression or original sin. In a moment of crisis the leader might even locate the enemy among his own followers. Such enemies generate opportunities for the charismatic personality to raise himself above others, and to present himself as the solution to some conflict or problem supposedly associated with these enemies. But the enemy also symbolises all those who once ignored or mocked the child narcissist, who blocked his path, including the primal other who opposed while also supporting him, the uncaring, unpredictable and potentially dangerous primary or secondary carer.

The charismatic personality is a good hater, which in itself suggests some frustration unresolved since childhood, but his hatred works for him. It stabilises him, giving him a focus and a target, and a definition of himself as the one opposed to such and such an evil. It also enables him to recruit others to his cause by using the threat of the enemy, and to enforce the compliance of those he has already recruited; in sum, he plays on their fears. The enemy symbolises objective reality to the charismatic personality, that ultimate indifferent other whom he is called to defeat, and all the smaller indifferent others whom he will destroy or escape at a time of his own choosing.

The sixth indicator of the false self is subtle and not often recognised. It can be approached via this question: In all the time that the charismatic personality is in the presence of another, how much of that time is spent in, and how much is spent out, of role? The answer is that most of them come to spend all of their time with others in role. They do it casually, freely, by choice and happily, but nevertheless they are acting a role and maintaining a position of authority or power over others. Why? The role is comforting and safe for him; predictable and ego-flattering. It is where he gets his narcissistic supply. What about time for relaxation? Time spent in role is relaxation for him and it is much more pleasurable than reading or swimming or walking alone. But what about those times when the work is done and everyone just sits around chatting, joking over a few drinks and telling their stories? Close observation reveals that even at such times the charismatic personality still conveys a sense of control and implicit leadership, if even indirectly from the rear and that he inserts a tone to the occasion that implies limits upon the actions and relationships present. Such moments may be time away from work, but they are not time out of role for him; he merely changes the form of his dominance.[1] The leader cannot abandon his role, and he must not. He is unable to just be real, to commune as an equal, and once an observer recognises this and watches the leader closely, it becomes quite obvious. As Australian Prime Minister Robert Menzies once said of Churchill, he was oratorical even in ordinary conversation (Baker, 2008).

The ability to commune is a real test of the charismatic personality. It is what he does least well because it requires him to suspend his leadership role, his grandiosity and his need for control, as well as all the tasks he habitually undertakes to keep himself occupied. It requires him to drop his guard and to allow the free play of the world and others around him, and to interact with the world and others from his natural being without trying to lead or influence. When he does get close to communing with others, his real self is most likely to surface with all its conflicts and hurts and fears, and this he avoids at all cost.

Lastly, there is surplus trauma, and a great underlying anger, usually inferred retrospectively from the leader's behaviour, especially the mess he leaves behind in his wake. For, despite whatever pain or suffering the leader may have experienced in childhood, he carries a much greater sense of grievance than is warranted. His subsequent feelings of outrage and entitlement endure for a lifetime, far longer than they would from any ordinary hurt that was gradually being processed and integrated. Kohut calls this 'narcissistic rage', a rage that shows by its extremity and longevity that some psychic wound has cut much more deeply than mere flesh and bone. Hitler is the obvious example; what possible childhood suffering could ever have justified his actions? He was however, quite comfortable with what he did; indeed, he continued to feel self-pity, sighing plaintively in his last days, 'If only I had someone to look after me'.

In Churchill's case he carried a lifetime sense of entitlement that was much more than merely the product of his class, and he clung to power as long as he was physically able. Churchill's frequent boorishness in his young adulthood suggests resentment towards the sensibilities of others, and his threat to 'break' anyone in his army unit who opposed him has to be viewed as indicating an underlying hostility towards anyone who refused his demands. He was opportunistic and impulsive to the point of deceit and disregard of others in some of his behaviours. He had the advantage of being on the side of right during World War II, but one of the traits of charismatic personalities is that they are attracted to problems where they can unleash their hostilities freely under the guise of being a warrior for the greater good. Churchill behaved well enough when things were going well, and his bad behaviour during crises is perhaps excusable, but in indifferent times he was often not a pleasant man to be around. He frequently behaved in an outrageous, pompous, bombastic and offensive manner.

Freud demonstrates this last point in an especially informative way. The ruthless ease with which he sought fame in disregard of the welfare of others and the protocols of science, the myth he spun of the opposition his discoveries provoked, his secretive control of the Vienna Psychoanalytic Society, and his jealous rivalries within it leading to his expulsion of dissenters, all suggest that he felt justified in taking whatever shortcuts and heavy-handed actions he needed to in order to achieve his goals. As he argued in his 1916 essay, some individuals feel that because they have suffered more than others, they are entitled to behave contrary to the ethics of their society, and without experiencing the pangs of conscience that others would. Beneath his compassionate demeanour and despite his basically gentle turn of mind, there existed a streak of hostility in Freud directed towards the world that he felt had made him suffer too much, and he was determined to overcome by fair means or foul whatever obstacles he encountered in his path.

The Charismatic Predicament

The charismatic predicament is another enduring theme that spans the entire lifetime, although it becomes increasingly important with age. It involves the same struggle everyone faces between the temptations of power and the need for love, a struggle that is all-pervasive, complex and deep, and that also involves the need for healing and integration on the one hand, versus short-term pleasure and egoism on the other. The stakes are higher for the charismatic personality because the power he wields makes the temptations of pleasure and egoism even more seductive; he really does have people calling him their saviour.

As numerous theorists have argued, there seems to exist an innate tendency in us all to strive toward psychological health in the fullest manner, to heal the wounds of childhood insofar as one is able, and to connect and bond with others in a realistic and loving way, while also attempting to achieve our fullest potential as individuals. This idea goes back to Aristotle and his notion of 'entelechy', according to which every individual sought to realise their own telos or destiny. More recently, Carl Jung posited the state of Individuation as the ultimate goal of psychological development, in which the person followed an inner voice, literally, a 'vocation', towards their psychological fulfilment (Jung, 1954). The Humanistic psychologists Abraham Maslow and Fritz Perls had versions of this with their notions of 'Self-actualisation' and 'Organismic Self-regulation' (Maslow, 1954; Perls et al., 1951). Carl Rogers also wrote of 'the inherent tendency of the organism to develop all its capacities in ways which serve to maintain or enhance the organism' (Rogers, 1959, p. 196). This idea is also present in Sartre's 'project', and in Heidegger's realisation of one's 'ownmost possibility', and in John Dewey's ethics of potentiality. Such striving was given a religious twist by Augustine when he wrote, 'The heart will never rest until it rests in thee' [God].

Heinz Kohut also had a version of this theory. He postulated a primitive entity called the 'nuclear self' that exists in the psyche in earliest infancy, and that drives development throughout life towards a goal something like self-fulfilment (Kohut, 1977). Kohut did not prescribe an ideal final state such as individuation or self-actualisation, saying merely that the nuclear self strives for expression, and that it carries an 'agenda'. This is a vaguely felt sense of emotional pushes and pulls that inclines one towards certain kinds of relationships that recapitulate and attempt to resolve one's earliest formative influences.

The thread connecting these ideas is that of a primordial or primitive state, an essential 'self' or even 'soul', that evolves through the vicissitudes of life. It eventually becomes — in what Kohut referred to as 'the curve of life' — an entity or mental state that is highly developed, refined and worldly, while somehow still remaining simple, elemental and natural. Thus it combines the wisdom and knowledge of life experience, with the innocence and purity of infancy.

Currently this theory is little more than a recurring motif that crops up in different forms in the social sciences when seen from different vantages, although there is accumulating and persuasive evidence that something like it occurs (Bohart & Tallman, 1999). The crucial addition of Kohut was to suggest that the final state attained by the highly developed self is in some way a recapitulation of an infantile, primitive state. This adds a cyclic, repetitive flavour to the theory that is reminiscent of Albert Camus's observation that headed chapter 3. Or, as T.S. Eliot famously wrote:

We shall not cease from exploration
And the end of all our exploring
Will be to arrive where we started
And know the place for the first time.

It is also generally agreed by most of these thinkers that the only thing that really fulfils is love; that the state which most humanises and heals is the state of loving subordination of one's own desires to the wellbeing of others. However, when this drive is blocked there is a tendency to fall back on short-term solutions and less fruitful satisfactions, in an effort to compensate for frustration and disappointment (Alderfer, 1972). Obvious examples include alcoholism and smoking among the disadvantaged and in lower socioeconomic groups, poker machine gambling among unhappy housewives, and obesity among television-dependent children. Basically what occurs is that the person attempts to reach for a creative, humane, deeply fulfilling solution to a problem, one that echoes his deepest needs, but finds he is unable to achieve it so he reverts to the consolations of impulse gratification.

For the charismatic personality this struggle is between the heady rewards of power that their abilities enable them to achieve, and the responsible use, or even the sacrifice of that power, for the sake of their loved-ones' wellbeing and greater humanity. The successful charismatic leader may find himself within easy grasp of all the temptations of flesh, wealth and power — an intoxicating brew — but not one that leads to fulfilment, love or personal growth. The great agenda of his creative project may satisfy in many ways, perhaps giving him an outlet for his disturbance and an opportunity to avenge his childhood suffering, as well as a way to achieve fame and fortune and to prove his superiority as a Great Man. He may genuinely achieve good works and attract much narcissistic supply. But such satisfactions still may not bring him closer to another. Nor do they automatically solve the significant personal and relationship problems that crop up in adult life, and which require compromise and negotiation among equals, an empathic understanding of others and a realistic appraisal of oneself.

As Masterson (1988) has argued, intimacy is the bête noire of narcissism and, by implication, love is his most difficult challenge. Unless the charismatic personality can invest himself in another and internalise this other within himself, he remains emotionally isolated, empty and frustrated, his needs to give and receive love remain unfulfilled, all sense of purpose and truth in relationships dies and his sense of connection with the rest of humanity diminishes to nothing. As Jesus put it best of all, 'What profit is it for a man to gain the world yet lose his soul?' Every parent knows what this question means, and even though we fall short of absolute yardsticks, nevertheless we know that there is no real alternative. This is the charismatic predicament: managing the pulls of love and power; the need for relationship versus the desire for strength and autonomy; commitment to a humble, fallible, needy, limited other versus the freedom and independence of wealth, fame and the delights of the senses. Wilhelm Reich, the clearest visionary of this fatal choice, has posed it as nothing less than 'The Murder of Christ', the decision made every moment of every hour of every day to choose between love and power (and like most prophets, he murdered his own Christ very successfully; Reich, 1963; Sharaf, 1983).

It probably begins with disillusionment. At some dimly felt level the charismatic personality probably guessed years ago (or maybe even knew all along), that fame and applause were shallowly based and fickle. Narcissistic supply satisfies today but fails to satisfy tomorrow, and as a highly intelligent person he must notice this, albeit he 'knows' that he is different from others. So when success fails to satisfy, he thinks that perhaps he needs to achieve more, and maybe he does, but that achievement in turn fails to satisfy for long. He turns this way and that, perhaps achieving every goal he sets for himself, but remaining restless and unfulfilled, bored and frustrated.

Actually, failure could be his salvation because it might force him to take stock and to examine himself more closely. Failure and loss are turning points for many people, enabling them to pause and reflect. But the charismatic personality who fails feels this as an offence to his being, and he redoubles his energy and commits to another project, angrily seeking redemption.

Ultimately, he is seeking love through his projects and he may even know this (Freud would have), or at least he knows the words and he has seen it in others. But to achieve realistic love he must abandon his fixation on power, his need to control and dominate, and his desire to be special. He must surrender, and in return he will be brought face-to-face with his vulnerabilities. He will have to culti-vate new and different skills. He will have to endure anxiety and the full neurotic force of another, and he will have no guarantee of success, for success in such matters is invariably only partial, and he will be left with regrets. He will find that unlike power, love is not simple and straightforward, but complex and fraught. Love is not something to be seized once and clung to forever; it is something that requires a renewed commitment each day. Such a commitment demands from him everything that he has, and perhaps even some things that of which he has no comprehension. The paradoxical logic of love, in which the more one gives away the more one has, eludes him because it is at odds with his zero-sum game mentality, and his win-at-all-costs impulses. To achieve realistic love, as distinct from narcissistic supply, he must learn much in painful faltering steps.

Much more than just romantic love is concerned here. He prob-ably finds being 'in love' easy. It enables him to momentarily escape from the prison of the self, allowing him to feel larger than he other-wise does, inflated and buoyed. Romantic love may function as yet another seductive and slippery form of narcissism in disguise. Unlike 'echoistic' love, genuine relational love depends upon the capacity of the lovers to desire and endure merger, while also being able to desire and endure separation and to maintain this stance over time. This is not easy. The temptation to opt for power even within a 'loving' rela-tionship, can blind one to the truth about one's relationship; that it may not really involve love at all. The Tibetan teacher Chogyam

Trungpa put it rather well when he advised, 'Ego can convert anything to its own purposes, even spirituality' (Trungpa, 1973, p. 55).

A recent book about Erik Erikson, written by his daughter Sue Erikson Bloland, describes what this can be like (Bloland, 2005). Erikson never knew his biological father; his Jewish mother had been abandoned soon after her marriage and later became pregnant to an unknown man. She was banished from her home in Copenhagen to Germany, where she had the baby alone, and she continued to live there during Erik's childhood. When Erik was three years old his mother remarried to a man named Theodor Homburger, and Erik was told that this man was his real father. At the age of eight he learned part of the truth, that Theodor was not his real father, but only much later still, in his adolescence, did he piece together the full facts of the matter, that there had been another mysterious man with whom his mother had relations outside of marriage, and this man was his father. All his life his mother refused to identify his real father.

Thus Erikson grew up feeling ashamed of his origins. By his own account he was something of a 'borderline' character in his youth and prone to depression. He was probably also fearful that perhaps, if his father had not wanted to recognise him as an infant, then maybe if he tracked him down and contacted him now, he might find that his father still may not want to accept him. Because of his Nordic appearance he was not fully accepted among his Jewish community, and because of anti-Semitism he was stigmatised in the broader German society. He seems to have remained ashamed of his origins into his old age, for even then he was still omitting significant details from his autobiographical writings.

To compensate, Erikson developed a sustaining fantasy that his father was a member of the Danish royalty, and that he too was destined for greatness. In time he met his future wife and they migrated to America in search of fame and fortune. After arriving they took new identities, ostensibly to give themselves new beginnings and to protect their children. He changed his surname to Erikson, signifying that from then on he was to be his own creation (Erik–Erik's son). His wife also changed her first name from Sally to Joan in an attempt to distance herself from her own painful childhood.

Erikson was powerfully intelligent and he developed great charisma. He studied psychoanalysis and pioneered child psychology, receiving posts at prestigious universities. All went well for many years, but the arrival of a Down syndrome child shocked him and his wife, and threatened their agenda. According to Bloland, their judgment failed them in this crisis because by then both had become highly dependent for their sense of wellbeing upon an idealised public image of success. They made poor decisions about this child and sent him off to an institution. In effect, they abandoned him because he was inconvenient to them; they opted out of responsible decision-making and selectively allowed others to tell them what they wanted to hear, thus justifying their own expediency.

Then came the great success of his masterwork *Childhood and Society* (1950), making Erikson famous as the formulator of an innovative theory of adult development, and the discoverer of the 'identity crisis'. In his daughter's words, 'I had never seen Dad enjoy himself so much'. He was awarded the Pulitzer Prize and received many other distinctions; there was even talk of a Nobel. However, celebrity did not heal his wounds; it only made it harder for him to confront his personal inadequacies. Fame compounded rather than healed his youthful traumas, because by then he had become heavily invested in his public image, especially the image of the benign and enlightened parent and educator, not the kind of person to feel crippling guilt, doubt, depression or anxiety, nor to consign a defective child to an institution. His fame, his charisma and his fawning audience became a trap for him, an idealised self that was also a false self, reflecting everything he yearned to be, yet utterly at odds with the reality of the never-healed lost child and the guilt-ridden adult. Indeed, he found that if he expressed his own frailties it actually enhanced his charisma because others took this as a charming affectation that just proved how secure he really was, that he felt no need to hide his vulnerabilities. In time, as Otto Fenichel has suggested, the love of the audience was 'needed in the same way as milk and affection are needed by the infant' (Fenichel, 1946, p. 148). He came to live a life of denial, a false self.

Towards the end of his life Erikson went into a psychological decline. He gradually withdrew socially and cognitively and his memory failed (though without the features of Alzheimer's). Eventually, he became virtually incommunicado and wholly dependent upon his wife. His daughter's theory is that he succumbed to the depression that had haunted him ever since childhood. The discoverer of the identity crisis knew full well the mismatch between his image and his reality, and the shame of his own double standard. His guilt over his rejection of his son weighed upon him, but he could not face his demons because to do so would shatter his carefully constructed self-image, and might cut off his flow of narcissistic supply.

Erikson played his charismatic role to the end, with tragic irony, for it was he who had described the charismatic personality as seeking to solve for others some problem that he had been unable to solve for himself. In Erikson's theory of adult development, the elderly person is faced with the alternatives of wisdom and despair, depending upon the integrity of how they have lived their life. In an interview that Allen Wheelis has described, a television interviewer asked of Erikson in his advanced old age, 'Have you achieved wisdom, Mr Erikson?' 'I'm afraid I have,' he answered (Wheelis, 1999, pp. 213–214).

Demise

Two scenarios may explain the demise of the charismatic personality. The first is based in classical psychoanalytic assumptions about power. It says that, at base, the charismatic personality feels vulnerable because he can never be secure in the power he has over his followers or their allegiance to him. Thus he is always anxious, and may be driven to ever-riskier agendas in order to enhance his power, yet is always left feeling unsatisfied and unfulfilled. This is the power scenario.

The second scenario is based on assumptions about love implicit in contemporary psychoanalytic thought and humanistic psychology. This position argues that what the charismatic personality really craves is not power but realistic love, a basic human need, and that the narcissistic supply that he receives always fails to satisfy this craving. His narcissism blocks him from feelings of vulnerability, and so attunes him to control and dominance that giving and receiving realistic love becomes impossible for him. Thus he demands ever more narcissistic supply from the followers, mistaking it for love and wondering why it leaves him unfulfilled. He may be driven to ever-riskier actions in order to get it, always hoping to fill his need but always failing because he is unable to recognise realistic love when it appears, or to internalise it when it is given. This is the love scenario.

Both scenarios probably occur to some degree. But it may also be possible that with very great narcissistic supply, and provided his childhood has contained at least some seed of realistic hope and fellowship, so that he is not entirely distrustful of the world and paranoid about others, and if life is kind to him, then he may be able to achieve and to settle for something 'good enough' in lieu of fulfilling his needs. If he can do this, then perhaps his plight may differ little from that of the rest of suffering humanity, those lesser heroes and heroines with their defeats, disappointments, alienations, renunciations and compromises.

To the external world looking on, and to history, the value and meaning of the charismatic leader resides in his mission and its success or failure. But from the perspective of a developmental theory of charisma, it is the final stage of life that is most revealing. It is here that an accounting occurs that determines whether the flight into charisma and the subsequent vicissitudes of that role have worked for the leader. This is to ask, how has he resolved his charismatic predicament? Has he achieved some peace within himself, in whatever way that he could, that has made it all worthwhile? Has he found some resolution of his conflicts and achieved something of lasting value that might ease whatever difficulties and disappointments his life has contained? How much does he have in his life that is real and nourishing?

It may be too much to expect a severely narcissistic person to 'only connect' in E.M. Forster's (2000) sense. His narcissism, which was once his refuge and his hope, thoroughly permeates his personality. This keeps him from deep intimacy and realistic love, though he may catch glimpses of these. But direct love expressed face-to-face is not the only kind of love. One may undertake great works as an indirect expression of love, thus attaining a more distant kind of connection that nevertheless sustains and redeems. This may be mostly just narcissistic supply (although not entirely), yet it may also reflect sincere and genuine caring.

Of the three charismatic personalities presented as case studies thus far, it seems that Churchill came the closest to achieving such healing and love. We cannot really know his emotional vicissitudes of course, but he gained the love of his nation; and in his 'dream' of meeting his father there is the healthy acceptance of a battle won, or at least fought to a draw. His marriage may not have been close, but perhaps it was as close as he could tolerate. He and Clementine shared the death of a daughter together, and he experienced the alienation of his oldest son Randolph. Of course, he clung to power and his leadership role as long as he could, but one need not be a renunciate to connect. His final pronouncement on life as a journey 'well worth making once', indicates both realistic ambivalence and affirmation.

The fate of Freud is perhaps less clear-cut. Clearly he saw what was involved, but he was caught in the grip of a very pessimistic mindset, a world-view not unlike Hitler's in some important respects, with its reduction of everything human to survival needs and its conflictual, rule of the jungle mentality. That Freud could feel deep compassion is shown by his lifelong liberalism and progressivism, by his humane relations with juniors (provided they supported his ideas), and by the general humanising thrust of psychoanalysis. But although Freud was a great intellectual, spiritually he was a small man who lost his faith in mankind, and he stayed firmly in control of his creation. Right up until his death he continued to promote himself through his ideas, and his sacrificing of his daughter Anna to the monument of his achievement, albeit he may have felt this to be the best outcome achievable, suggests a despairing heart. We may suspect that Freud went to his death stoically unresolved.

Hitler represents the ultimate instance of a charismatic personality too damaged for redemption. To Hitler, vulnerability equalled weakness and this threatened him. He was unable to invest himself in others or to internalise them within himself; to deeply and genuinely care for another. He embraced the might is right principle as few other leaders in history have done, but it left him no room to grow as a human being. So total was his commitment to power that he steadfastly refused to mature or develop emotionally. It was a point of pride to him that he changed little over the years, in defiance of age and external events. He remained defiant right up until the end, prepared to destroy the German people as well as himself, rather than consider that he might just be wrong.

A test of the wisdom of any older person is the degree of acceptance they have of themselves and their mortality. This means graceful acceptance of the limits of one's achievement, a recognition that today's truth is invariably superseded by tomorrow's, that the future belongs to the young, and that whatever one's life has meant to one it will soon mean much less to those one leaves behind. This constitutes a severe narcissistic wound to the charismatic personality, one that he may refuse to accept, angrily lashing out at others who invariably disappoint and desperately clutching at the tokens of power ever more tightly as time erodes his real power. As he watches his power slipping he may be tempted to up the ante, to pursue ever more extreme agendas in a futile effort to deny his impending demise, and such efforts will be all the more desperate the less genuine love and achievement he has in his life. With life growing shorter he may start to think something like, 'I have only so much time in which to achieve my life's work, I must move faster'. Thus he may impulsively endanger himself and his followers, whom he may come to care for less and less.

If he has not achieved some justification in his life, if he has found little that is nourishing and real despite whatever successes he has achieved, and whatever narcissistic supply he can command, then he will gradually come to feel emptier and emptier. He may be driven to ever more desperate or even antisocial efforts to gain love, although only able to tolerate it vicariously as narcissistic supply. He may cling to the reins of power as long as his weakening fingers can grasp. Certainly, he will not plan a succession unless he feels fairly secure, and he has some acceptance of his place in history and its limitations. Nor will he bestow his legacy in a rational way upon those reasonable supporters and sympathisers who have remained realistic and honourable. Instead he may fling it to the winds, squandering it randomly on those whose relationships with him have been opportunistic or perverse or sycophantic, in a final act of revenge on those who have proven unworthy. Or he might revert to the lesser satisfactions that he is able to achieve, unable to solve his charismatic predicament, unwilling to desist and eventually uncaring of whom he hurts and how he lives. This is the recipe for charismatic failure, and is a common enough fate. Still, it may be instructive to compare the frequency of charismatic failure with the incidence of flawed leadership generally, which has been estimated at between 60 and 76 per cent (Hogan, Raskin, & Fazzini, 1990). The failure rate of the most senior leaders in corporate America — presumably the best of the best — is about 50 per cent (van Vugt, 2008, p. 45).

We now know what it means to age well. Science may be value-neutral but human beings are not, and evolution has given us a bias towards affiliation and prosocial behaviours, towards nurturing and connection. When all the allowances have been made for eccentrics, geniuses, artists and extraordinary individuals, as well as all those others — movie stars, activists, soldiers of fortune and the rest — whose life situations deviate markedly from the norm, nevertheless, even after all these exceptional people have been considered, a remarkably homogenous picture of the optimum human life course emerges. The key is generativity (Vaillant, 2002); nothing enriches quite like looking after the next generation, and charismatic personalities are no different from the rest of us in this. To invest one's substance in forms of life and work that will outlive one is ultimately what brings the greatest warmth to one's declining years. This is about having some purpose or duty or ethic that one recognises as in some way 'above' one; having something to aspire to that transcends one, and thus to serve an ideal that one feels one belongs to, and is nourished or even 'saved' by.

One may engage in generativity without connecting directly to others, and this seems to be the optimal fate for charismatic personalities. Denied intimate connection by their narcissism, they may still guide the young or their community or nation in a benign way that brings them great comfort. Churchill and Freud both did this. However, we die as we live. Already distorted by early life influences, the narcissistic personality may suffer additional life stresses that pervert them further. The last shreds of their connections to others may be broken by such experiences, leaving them emotionally crippled. Such a person may become embittered and permanently regressed to primitive impulse gratifications (Alderfer, 1972), reduced to a mere function, albeit a dominant and antisocial function. Perhaps the best illustration of this involves Mao Tse-tung.

It is wonderful how historians and biographers can bypass the significance of something so fundamental as that Mao was the first of his mother's children to survive infancy, and that she lost four others during or prior to his early childhood. If she was typical of most peasant women (and there is no reason to think she was not), this experience would have traumatised her deeply. Thus she would have anxiously invested most of her hopes in her remaining children, especially the first to survive, a boy — little Mao. Biographer Philip Short's concession that she 'may have had dreams for him' is surely the minimum that can be surmised (Short, 1999, p. 23).

Mao was born in 1893 into a family of wealthy peasants; like few Chinese children, he had his own room and books. He was surrounded by other supportive women — aunts and a grandmother who treated him as their own child — and he came to worship his mother. For some reason his mother preferred to live with her own family (she eventually left her husband) and Mao came to hate his father, who nevertheless paid for his education and eventual training to become a teacher. His father betrothed him at 14 to a girl several years older than he, but he walked out without consummating the marriage.

As a youth, Mao was a brilliant student who read voraciously. He was considered charming and intelligent by teachers and classmates. He wrote poetry (including love poetry) and had an excellent memory; he was funny and he had an eye for the ladies. His thinking impressed others as unusual and original, and he used vivid, colourful language. However, he was also seen as arrogant and a bully, and privately he struggled with self-doubt and was prone to depression. He had no close friends and he was thought to be something of an actor. He had unbounded self-confidence, yet he could be curiously ill at ease with strangers. He was highly anxious and he suffered psychosomatic illnesses that sometimes kept him in bed for a week or more; he had several 'nervous breakdowns' over the years. At such times he might lapse into self-pitying melodrama; throughout his life he relied upon a variety of medications.

Mao's narcissism was evident early. He was intensely opposi-tional, often grandiose, could not admit error, had very little empathy and he craved acclaim. He had tremendous energy (with associated sleep problems), and he was astonishingly prescient, sometimes being the only one in his circle who could accurately predict what might happen in a given situation. His powerful will was noted by others from his earliest days, and it gave him an elevated, inscrutable air; one commentator wrote of his 'deep caverns of spiritual seclusion', and of 'a door to his being that had never been opened to anyone' (Short, 1999, p. 2). He came to see himself as a 'Great Hero' entitled to wage 'giant wars' to destroy and reform China. He wrote:

> I do not agree with the view that to be moral, the motive of one's action has to be benefiting others. Morality does not have to be defined in relation to others ... People like me want to ... satisfy our hearts to the full, and in doing so we automatically have the most valuable moral codes. Of course there are people and objects in the world, but they are all there only for me ... People like me have only a duty to ourselves; we have no duty to other people ... I am concerned only about developing myself ... I have my desire and act on it. I am responsible to no one. (Chang & Halliday, 2005, p. 13)

Mao came of age during a time of breakdown in the central authority of China, a period when warlords ruled by ad hoc violence and shift-ing coalitions. He searched long and hard for an 'ism' to coalesce his troubled personality around, and finally settled on Marxism in his mid-30s, becoming a revolutionary. But the revolutionary war became one of massacre and counter-massacre in which the most effective tyrant prevailed, and this experience corrupted him. Mao initially rejected the use of violence. He converted to it in the face of the slaughters perpetrated by the warlords, and by Nationalist leader Chiang Kai-Shek and later still by the invading Japanese. He was also influenced in this by the numerous bloody precedents of history (life was cheap in old China), and because his Russian communist patrons were heading down the same path. From being an idealistic youth sympathetic to women and the plight of the poor, in time he became increasingly murderous, paranoid and devious. The momentous events he was involved in consumed him and his humanity 'withered on the vine' (Short, 1999, p. 317).

During the revolutionary struggle Mao's brilliance found its niche. Through firsthand experience in the field he taught himself and eventually mastered disciplines as diverse as military strategy, organisational management, leadership, international diplomacy and even killing and interrogation. He issued personal instructions on the conduct of torture, leading some recent biographers to write that 'Mao's most formidable weapon was his pitilessness' (Chang & Halliday, 1999, p. 324). Along the way he looted riches for himself and traded in opium. He also became a highly effective manipulator, but he paid a price; even when he was relaxing he was always on his guard (Li, 1996). He also lost many loved ones to the struggle, including several of his own children, and this must have taken an almost unimaginable toll on him, but he pressed on with his great cause.

Mao emerged victorious but success and adulation inflated him (Li, 1996), and his disappointment in marriage led to him increasingly turning to the sexual services of young women for consolation. By the time he met Nikita Khrushchev (who was horrified at his megalomania) he had become something of a monster. Another biographer has written:

> In the mid-1950s Mao was so convinced of the essential correctness of his own thought that he could no longer comprehend why, if people had the freedom to think for themselves, they would think what they wanted, not what he wanted. (Short, 1999, pp. 470–471)

As ruler of China, Mao initiated many eccentric policies, some of which resulted in great suffering to the people, while others were more successful. During his tenure life expectancy in China doubled, literacy rose from less than 20% to over 90%, and the population almost doubled. But the failure of his Great Leap Forward, which involved fast-tracking industrialisation, forced collectivisation and the abandonment of birth control, seemed to cause Mao a deep narcissistic injury. He refused to believe that his policies had been so mistaken as to cause the death by starvation of about 25 million people. By now he was not much troubled by the deaths in themselves, for he had long been talking in terms of losing a few million or so here or there. It was the reproach of those around him that most offended him.

Partly in rage, he launched the Cultural Revolution to punish ungrateful humanity. By the end of these campaigns the population of China was thoroughly disillusioned with him, the second-tier of leadership, intellectuals and the civil service were utterly traumatised and the economy teetered on the brink of collapse. In rural areas banditry reappeared, and he was denounced (very cautiously) by even his most trusted lieutenants. When Lin Biao attempted a coup, Mao felt the betrayal personally, for Lin had been one of his longest-standing comrades. But it was not the coup attempt so much as Lin's escape and his refusal to return and abase himself that most distressed him (Lin actually died in a plane crash while fleeing). Mao went to bed depressed and remained there nearly two months; he never regained his old vigour. He too considered himself a failure.

Mao's last years were squalid and despairing. Utterly paranoid and reduced by weight loss to a shadow of his former self, partly paralysed and with several illnesses, including heart disease, lung disease and terminal Lou Gehrig's disease (and a lesser venereal disease that he refused to have treated), incontinent and drooling, his speech guttural and incomprehensible even to close associates, almost blind and enduring bedsores, he now needed oxygen tanks and his doctor to accompany him even on short journeys of a few hundred metres. Impotent (he had long been sterile), he was now beyond even the consolations of his groupies, although one remained loyal to the death. Yet he remained lucid to the end, and he had one more trump to play.

If he wanted to die in his bed, Mao needed a protector at this extremity, so he brought his wife Jiang Ching in as his new, most trusted lieutenant. Jiang had long hungered for power, and she knew that any power she had derived from him, so it was in her interest to keep him healthy for as long as possible. Despite her longstanding revolutionary credentials, Mao had kept her out of the Politburo till then because he had come to loathe her, but now they needed each other. Jiang had a well-developed paranoid streak and could be utterly ruthless, so she served his purpose well. Mao had no genuine concern for her however, and he privately instructed his generals to postpone any coup until after he was dead, when they could do whatever they liked with her.

To almost everyone Mao touched he brought misery and death. Biographers Chang and Halliday calculate that he caused the deaths of about 75 million people, perhaps a high estimate, but even more sympathetic commentators accept a figure roughly equivalent to, or just below, that of the entire death toll of World War II. A staggering number of his political associates — revolutionary comrades whom he had fought alongside — were 'purged' by him, including almost his entire inner circle, as he cut down every possible rival or successor. Of his wives, the first suffered the disgrace of his rejection and died an early death. He abandoned his second wife, Yang Kai-hui, and refused to rescue her and their children from the Nationalist forces; she and Mao's adoptive sister were executed by them. He also abandoned his third wife, He Zizhen Gui-yuan, and their children, several of whom were given away at moments of crisis and never heard from again. She had suffered greatly from the revolutionary struggle, and she ended her days broken-hearted and needing ongoing psychiatric care.

Mao had nine children but he became indifferent to most of them; only the death of his oldest son, Anying, during the Korean War caused him significant grief. His other surviving son, Anqing, was schizophrenic. Mao's youngest brother, Zetan, died in a clash with the Nationalists, and his second brother, Zemin, was killed by a warlord. His two surviving daughters sided with their mother Jiang Ching against him; one was mentally unstable.

How could Mao do the things he did (or, as he would have preferred it, been obliged to do)? He could not have foreseen, before he began his charismatic adventure, where it would lead him. Despite his brash and youthful confidence he began as only an idealistic and restless dreamer, whose narcissism might have healed or at least moderated over time. But the lightness of his intimate connections made him perfect for the sacrifices that would be needed in the revolutionary struggle, while also inclining him towards such a mission. Still, he must have felt increasingly strange as his intimate connections fell away, and as he ventured further out into unknown regions of barbarity and expediency for an ideal that became increasingly costly. This too may have served the needs of the times, for only someone who had already lost their own loved ones would have felt entitled to continue to inflict such losses on others. He too saw himself as an 'exception'.

Mao's actions would always have seemed justifiable and necessary, at least by his lights. But as his losses mounted and his youthful image of himself as a Great Man of Destiny was overtaken by grief, isolation and bloodshed, he would have felt increasingly lost, confused and enraged, until eventually he may have just switched off his remaining moral senses. He probably always hoped for some redeeming comfort to dawn after the next great hurdle had been crossed. Eventually, it may have seemed that the only thing that might redeem or justify his losses, or at least make them seem tolerable, would have been ever-greater achievements leading to eventual world domination which, it seems, he did dream of. But in reality his achievements brought less and less satisfaction, while growing ever more costly. When his last intimate connections were severed there was no way back. He was inclined irrevocably on a path towards the eventual self-destruction of his humanity, becoming himself the oppressor that he had once despised.

There is probably only one interpretation of promiscuity into one's old age; failure to connect. According to his doctor, Mao's sexual tastes became more extreme as he got older, but nevertheless Jiang Ching performed her protective function right up to the last. There was no longer any love in their relationship, and when she was duly arrested four weeks after he died she went quietly, knowing that she had no power base without him. In 1991, she committed suicide in the prison cell to which the plotters whom Mao had instructed to delay their coup had consigned her.

Endnote

1   This might be seen as the charismatic personality maintaining his false self as an 'ideal self' in Karen Horney's sense, and this would explain why it feels so natural and comfortable to him. Horney advises:

> In contrast to authentic ideals, the idealized image has a static quality. It is not a goal towards whose attainment he strives but a fixed idea, which he worships ... The idealized image is a decided hindrance to growth because it either denies shortcoming or merely condemns them. Genuine ideals make for humility, the idealized image for arrogance. (Horney, 1945, p. 99)

• • • • •

PART TWO

An Exploration of Charismatic Forms

CHAPTER NINE

The Charismatic Relationship

The hidden purpose of the charismatic group is not to succeed but to experience itself.

Charles Lindholm

In Max Weber's original formulation, charisma was not just a personal quality such as leadership ability or strategic vision or divine inspiration that inhered in a person. It also depended upon how the leader was perceived by others, who believed that he or she possessed such qualities. According to Weber, charisma was a quality of a relationship, and he defined it as:

> ... a certain quality of an individual personality by virtue of which he is considered extraordinary and treated as endowed with supernatural, superhuman, or at least specifically exceptional powers ... [that] are regarded as of divine origin. (Weber, 1968, pp. 241–242)

The approach taken in this book is to assume that while charisma is a quality of a relationship (as Weber argued), nevertheless certain persons are much more likely to be regarded as charismatic than others, specifically those who combine narcissism (as Heinz Kohut suggested) with exceptional talents such as creativity, rhetorical skill and so forth (as Weber also allowed). Hence, it is possible to speak of the charismatic personality as one who is most likely to be viewed by others as charismatic because of their talents and disposition. However, some account must also be provided of quite why the followers — and not everyone — perceive the leader thus, and this requires discussion of the charismatic relationship, and the natures of the followers.

To call someone a 'follower' can be something of an affront, suggesting a person who has no will of his or her own, so to avoid offence another word is often substituted. The inspiring new CEO brought in to revivify a moribund business does not have followers, he has associates and employees. The dynamic head of department of a progressive academic college does not have followers, she has colleagues. Vladimir Lenin did not have followers, he had comrades; Bill Clinton did not have followers, he had supporters; and Elvis did not have followers, he had fans. Then there are sympathisers, adherents, helpers, backers, disciples, admirers, acolytes, allies, defenders, benefactors, auxiliaries, sidekicks, assenters, friends, patrons, initiates, participants and numerous other designations. But regardless of whatever word is used, all these relationships involve leaders and followers. The choice of descriptor is largely determined by whatever language is considered acceptable in a given context. We might even suspect that spiritual and social movements who freely accept the term 'follower' are actually being more honest about the relationships involved. For present purposes, all those who freely support a charismatic leader from a stance of affectionate hope, and who demonstrate energised commitment and idealisation, shall be considered 'followers'.

The psychology of the follower is not at all like the stereotype of the brainwashed fanatic or devotee portrayed in the media. Beneath appearances, four crucial factors are involved — faith, trust, courage and projection (Kohut, 1985; Smith, 1962, 1979). First, the potential follower must have basic 'normal' mental functioning, including some seed of faith that creative personal change is possible, and a hunger for advancement or even transcendence, however these are viewed. Many people seem not to have such faith; they are happy to live relatively comfortable lives without reflecting too much or taking risks. Nor do they yearn for what Dietrich Bonhoeffer (1995, p. 444) called 'the beyond in the midst of life' because they are simply unaware of their depths. As Joseph Conrad has Marlow say disparagingly in *Heart of Darkness*, 'A fool is always safe'. Charismatic groups may attract many damaged people who are accepted on compassionate grounds, but they do not form the core, nor are they among the active drivers of the movement.

Second, some ability to trust in a fairly profound way is also required, and again not everyone has this capacity. As psychotherapists know, many people are unable to benefit from therapy because they cannot allow themselves to expose their needs, or to trust another with their vulnerabilities or the details of their personal lives. The quality of trust required to follow a charismatic leader can amount to entrusting one's life to another, and there has to be some basic belief in the fundamental trustworthiness of others, and also of oneself, some modicum of personal security, in order for this level of trust to exist. The British people during World War II came to believe that Churchill would rather die himself than surrender to the Nazis, and indeed he may have. But despite the crisis they were facing, their trust was an outgrowth of strength and health rather than of weakness or psychopathology, and it presupposed a realistic knowledge of human nature and its capacities and limitations. Such trust is never given arbitrarily.

The third factor is courage. The follower risks going outside his or her comfort zone of familiar experience to enter a charismatic relationship, knowing that from then on life will be very different and unpredictable. They can safely assume that friends and family will think them foolish, and that outsiders and the media may ridicule them. One follower of the Hindu guru Rajneesh likened this to being prepared to step off a precipice into nothing (Gordon, 1987). Of course there is need, but probably more creative need than sheer destitution. As the account of Freud and Fliess shows, Freud was attempting to venture far beyond where other thinkers had gone, into a lonely realm of possible chaos and madness. Charisma is not for the faint-hearted.

The fourth factor is projection. When faith, trust and courage come together, the follower begins to project onto the leader the qualities that he or she needs to see in them in order to undertake a major creative effort. The follower comes to view the leader as embodying some elevated or divine state, perhaps as a visionary or an inspired prophet, the voice of God or at least of the future. There is idealisation and love, sometimes worship even, and the follower comes to feel more complete and whole through identification with the leader. Such idealisation may be partly misplaced naiveté, and it may also indicate that the follower has been duped, but at base it is a willed act of surrender and a courageous investment of great hope in the leader.

The follower dares to believe that the leader has the key to his (the follower's) future, and willingly enters the charismatic relationship in a spirit of loving commitment. At its healthiest, this is really a commitment to the values and ideals that the leader represents. But not all of the followers think in such abstractions. Some prefer to commit to the person of the leader, identifying their fate with his. This is a problematic stance because it means that the relationship loses its self-correcting capacity; the welfare of the leader becomes all important, and reality-testing and self-checking are diminished. If the follower commits to the ideals that the leader espouses and then the leader strays from these ideals, the follower will feel free to desert him, but if the commitment is to the leader's person then there can be no such falsification or disillusionment. In most charismatic groups there is a small inner elite who commit to the leader's person, and they usually come to occupy the upper echelon of participants; they pose a grave danger.

Faith and trust provide the background for the charismatic relationship; without them it could not occur, but they are the passive side of the follower's affections and are evoked by the leader. The active side involves the courage the follower summons up to make his commitment to the leader, and the projections that the follower makes onto the leader in order to see him or her in the way best suited to the follower's aspirations.[1] This 'surrender in the service of the ego' (Oakes, 1997, p. 137) appears to be regressive because the follower imbues the leader with parental elements, and indeed, there are invariably some regressive aspects to it such as feelings of awe, dependency and obedience. But it is freely willed and somewhat controlled; the follower has not lost his or her self, nor been brainwashed or hypnotised, although perhaps swept up in a wave of enthusiasm. The follower sees what he needs to see and feels what he needs to feel in order to generate the right quality of hope and inspiration that enables him to fulfil his creative agenda. In sum, the follower's unconscious projects into his consciousness whatever resources are needed to undertake a creative challenge.

Such creative regression may seem pathological, but it is not. Creativity usually happens behind closed doors, in artists' studios and the like, but the charismatic follower has no such refuge. Charisma is one of just a few experiences in which the unconscious mind appears to overtake consciousness and produce extraordinary passions and seemingly irrational behaviour. Romantic love, creativity and mysticism are similar in that they also flood consciousness with unconscious contents. In romantic love, an individual who objectively seems similar to everyone else suddenly becomes Mr or Ms Right, as the lover engages with their unconscious needs. In art, the artist who is overtaken with creative inspiration may refuse to eat or sleep until their project is completed. Similarly the mystic, seized with the rapture of God, giggles stupidly and weeps unrestrainedly, and when asked why, cannot explain. We recognise that these people are not psychotic, but that something very strange is going on in them, as some process from beyond the conscious mind infatuates, inspires, or enraptures them. Something similar occurs with charisma.

Kohut was at pains to explain such apparently pathological and irrational regression. In an essay titled 'On Courage', which he suppressed until after his death and that was only published in 1985, he considered the mix of regression and projection that occur in acts of great courage. He studied the writings of lone German resisters of the Nazis and found that prior to them taking their moral stands, stands that they knew would probably cost them their lives, many underwent seemingly pathological experiences such as prophetic dreams and full-blown hallucinations in which they felt that God spoke to them, or that some powerfully symbolic message had been imparted to them. Yet these people were not psychotic; rather, they were among the emotionally healthiest, most ethical people in the German nation at that time. They needed support during a life and death moral crisis so they created an image of a God-like figure — presumably a projection from their unconscious — from which to draw strength and inspiration.

Kohut argued that the ability to create, in extreme situations, the fantasy of being supported by a God-like omnipotent figure, or some other symbol of moral purity, is one of the assets of the healthy personality, despite its apparently pathological features (Kohut, 1977, 1985; La Barre, 1980). The significant difference between the fantasy projections of creative heroes and the delusions of disturbed individuals lay in the peace that creative resolution brought with it. An example of this would be Dietrich Bonhoeffer's calm during his imprisonment, and especially in the time leading up to the last days of his life before his execution by the Nazis. Other prisoners took heart from his serenity, and he was able to comfort them appropriately because of the inner peace that his struggle had produced (Bonhoeffer, 1997). Kohut argued that the consummate peace achieved by the hero, enabling him or her to even face death calmly, comes from fulfilment of an agenda laid down in the nuclear self during infancy, and purposefully pursued thereafter (Kohut, 1985). This is the psychological meaning of saving one's soul while losing one's life.

The stereotype of pathology associated with charismatic followers has also been falsified by research in organisational psychology, which has shown the many benefits of charisma to the followers. In workplaces, charisma motivates employees to make greater efforts, to behave better, and to engage in fewer deviant behaviours.[2] Groups led by charismatic leaders are more satisfied than comparable noncharismatic groups; they have lower turnover rates, are better able to deal with frustration and experience an enhanced sense of optimism.[3] Charismatic followers have higher levels of creativity, are more positively entrepreneurial, and experience less stress than other employees despite their more energised commitment (Bass & Riggio, 2006, pp. 37, 54; Elkins & Keller, 2003; Jung, 2001). In a fascinating study comparing the outputs of workers under charismatic and noncharismatic leaders, it was found that while noncharismatic leaders elicited greater quantitative performance from their workers, charismatic leaders elicited greater qualitative performance; the authors suggested that trust mediated the relationship between charismatic leadership and quality of work (Hoyt & Blascovitch, 2003). There have also been significant mental health gains shown to occur among members of charismatic religious and communal movements (Kilbourne & Richardson, 1980; Richardson, 1995).

Given these follower characteristics, some have argued that charismatic leaders are 'constructed' by their followers, who thus are really the true leaders, being followers in name only and using the 'leader' for their own agenda. This 'follower-centred' theory, most closely identified with James Meindl (1995), resonates with those unsympathetic critics who, for various reasons, refuse to see anything extraordinary in the charismatic leader, and express amazement at how easily people can be led astray by deluded, Svengali-type figures. The assumption (not Meindl's, for he was much more sophisticated), is that only the weak-minded could fall so easily under the spell of another, therefore there must be another explanation for the strong-willed followers. These critics virtually define charisma as an ability to brainwash and mesmerise vulnerable followers; strong followers 'construct' such a leader for their own purposes.

While some followers really are dependent and easily led, the majority are not. Indeed, many may have previously been quite successful and independent in their lives before joining the leader. Charisma also tends to occur more among the bright and able than among the downtrodden or simple-minded. The critics of charismatic relationships usually either have quite different values and thus are impervious to the appeal of the leader, or they are those who stand to lose the most should the charismatic movement be successful, and thus they are at pains to deny anything of value about the leader. Actually, the follower-centred approach contradicts those who would dismiss charisma as a popular delusion for it implies that the followers are rational goal-seeking actors (Shamir et al., 2006).

Finally, to suggest that the followers are deluding themselves and that charisma exists only in the eye of the beholder, is to ignore the almost physical manifestations of charisma that are attested to by so many; typically, the 'glow' surrounding former Australian Prime Minister, Gough Whitlam, or the power 'streaming' from the shaman, or the 'piercing' eyes of Adolf Hitler (La Barre, 1980; Waite, 1977; Walter, 1985). Of course, these descriptions are metaphors but their intent is to describe the very physicality of charisma. The magnetic eyes of charismatic personalities have been remarked upon by so many credible independent observers that any suggestion that it arises from the subjectivity of the followers is unsustainable (Oakes, 1997, p. 20). Perhaps the most extreme example of the 'objectivity' of charisma was George Washington (Flexner, 1974). According to virtually all reports, even as a youth he was so naturally distinguished and possessed such a commanding imperiousness as to impress all who met him, even those who opposed him. He maintained this magisterial demeanour even informally when relaxing with others just playing cards, for although it seems that he was somewhat stuffy and formal, he nevertheless did enjoy spending time with associates joking, drinking and talking. He could seemingly be at one and the same time both elevated and just one of the men. So inevitably stately, so comfortably noble was his carriage and bearing, that numerous superlatives flowed his way. It has been said that he exuded 'masculine power', and that he possessed 'martial dignity'. Marquis de Lafayette, describing how Washington turned retreating troops around to face and rout the enemy, testified:

> General Washington seemed to arrest fortune with one glance ... His presence stopped the retreat ... His graceful bearing on horseback, his calm and deportment which still retained a trace of displeasure ... were all calculated to inspire the highest degree of enthusiasm ... I thought then as now that I had never beheld so superb a man. (Flexner, 1974, p. 123)

The effect of Washington on his troops has been described by Ludwig van Closen who wrote:

> It is incredible that soldiers composed of every age, even of children of fifteen, of whites and blacks, almost naked, unpaid, and rather poorly fed, can march so well and stand fire so steadfastly ... (because of) the calm and calculated measures of George Washington, in whom I daily discover some new and eminent qualities ... He is certainly admirable as the leader of his army, in which everyone regards him as his father and friend. (Flexner, 1974, p. 157)

Yet another observer, Tobias Lear, who worked for Washington as his secretary, after a period of mutual testing and reserve, wrote that he had had:

> Occasion to be with him in every situation in which a man is placed in his family — have ate and drank with him constantly, and almost every evening have played at cards with him, and I declare that I have never found a single thing that could lessen my respect for him. A complete knowledge of his honesty, uprightness, and candour in all his private transactions have sometimes led me to think him more than a man. (Flexner, 1974, p. 186)

In sum, there are genuine heroes, and to suggest that charismatic followers are merely constructing such a leader to suit their own irrational needs flies in the face of the facts. Of course, not all charismatic personalities are as virtuous as their followers believe them to be, perhaps only a few are; certainly Washington behaved poorly at times, spending unwisely, and inappropriately gaining land in areas beyond the law (Flexner, 1974, pp. 46, 55). But charisma is real enough; that is, it exists objectively — almost physically — beyond the eyes of the beholders (needy or otherwise).

Like all relationships that are personal, dynamic and growthful, typically family and healing relationships, the charismatic relationship is multilayered; any summary of it is bound to be incomplete. The varieties of charismatic experience are as diverse as the people involved. However, because of the interactions of several constants that occur in virtually all charismatic relationships, there are only a few main pathways and these are mostly quite predictable. Hence it is possible to describe in outline the central dynamics that define the charismatic relationship, at least in most cases, most of the time. This chapter discusses five key constants: the follower's Great Work, the evolving relationship between the leader and follower, the Intimidating Effect, the Charismatic Crucible, and the Charismatic Moment.

The Great Work
The charismatic relationship begins when the needs of the follower mesh with the needs of the leader. They need each other, but for different purposes. Beneath the explicit purpose of the relationship —

typically an employer–employee relationship, or a leader–follower relationship, or a guru–devotee relationship — the follower hopes for something much more that invariably involves personal change. This constitutes a Great Work that he or she hopes to achieve, an objective goal that is not always consciously stated, but is expected to bring happiness and even transformation, salvation, enlightenment, unconditional love, or something else of at least this order. To the follower the relationship feels special. It is about progressing, going forward to become something better as he identifies with the leader and emulates him.

The leader also has a twofold relationship. Superficially, he sees the follower as able to play an important role in his agenda, to help grow the church or the business or the movement. But at a deeper level the leader needs the follower emotionally in order to feel that which he 'knows' he already is — elevated, special and extraordinary. Hence the leader does not hope to personally grow from the relationship or to spiritually progress (or at least, not in any sense that is even remotely like what the follower might define as spiritual progress). To the leader the relationship is about regressing, going backward to an archaic state that he believes is his entitlement, and which has shaped the core of his being. This is why charismatic relationships are time-limited. The leader and follower are going in different directions emotionally, and the follower is destined to outgrow the leader. Hence, the follower has a limited time to get the best out of the relationship, after which he becomes disillusioned and the relationship becomes toxic.

The follower joins the leader to further the leader's agenda because it fits in with his own deeper needs; it resonates with his own great work. He is not passively recruited. Both leader and follower find each other for their own purposes (Little, 1985). In organisational psychology this is called a 'transformational' relationship, as opposed to a mere 'transactional' relationship in which only money, goods or services are transferred. The transformational relationship is characterised by an energised commitment on the part of the follower, by deep affection towards the leader and an idealisation of him, and by a powerful sense of shared purpose and hope for better things to come, of destiny even. It involves ultimate concerns in some sense, perhaps as salvation in this world or the next, or in some private internal communion such as when the follower feels that the leader speaks just to him, or is able to read his heart like a book as only God or a prophet — or at least someone extraordinary — might.

The follower's great work may be something such as working with and growing within an organisation that he or she believes will make a positive difference to the world. Typically, the relationship is between the exciting new leader and his keenest followers, employees or supporters in a company, party or movement. Or, it may involve the follower joining a spiritual elect in order to marry and have a family in true accordance with God's divine plan (as revealed by the leader). Or, it may involve membership in a political party that aims to revolutionise society. As the Churchill example demonstrates, the followers may be seeking a saviour in a time of national crisis, and the charismatic relationship may thus occur on a nationwide basis. As the Freud example shows, the creative person may need emotional support during a creative crisis, and the relationship may be informal with no institutional framework. No doubt there are other variants also, but whatever the specific context there is usually some feeling of crisis, or at least urgency, and of great matters at stake. In the early years the follower can be counted on to hold nothing of himself back from the stated program, for in advancing it he believes that he is advancing his own great work.

Then there is the Pygmalion Effect. The leader instils in his followers the idea that they can perform up to a very high level, raising their self-esteem and assuring them that the leader will be on hand to help them (Eden & Sulimani, 2002). Sheer belief in one, from such an idealised elder, can produce magical effects and the leader adroitly administers such favours among the followers whenever the opportunity arises. In following the leader, the followers are gratified to discover strengths within themselves that they never imagined they possessed, and this leads to renewed commitment.

The success or failure of the follower's great work may take a short or a long time depending on the nature of the work, but at some point he either achieves what he came for or he does not. In a typical instance, the worker who joins a charismatic company to grow personally and to improve the world by following the mission of a visionary entrepreneur feels that he has grown greatly at a personal level and contributed his piece to worldly salvation. Or the follower who joined a spiritual group to marry and have a family in 'God's community' does indeed marry and have her family. Or, the charismatic politician wins office and the supporter is given a useful job in government and sets about helping to reform the country.

Each of these scenarios may lead to the follower reassessing their commitment, not because something has gone wrong, but simply because she or he has achieved what was sought, and now — in Thoreau's words — has 'many more lives to live'. There are several options. The follower may set new goals that do not require a charismatic relationship, and which may in time lead to conflict with the leader. Or they may leave immediately if there has been some backlog of tensions. Or they may stay for a while because it is comfortable and then leave when some 'crisis' occurs (charismatic groups have many 'crises,' both real and contrived). Or the follower may leave when the charismatic leader espouses a new agenda that does not appeal. The leader will invariably resent such leaving, believing that the follower still has much to achieve under his direction, and many a follower has announced their intended departure expecting to be well-wished on their way, only to arouse vitriolic anger in the leader. The leader does not like to feel abandoned; it outrages him by showing that others are not really 'parts' in his narcissistic world.

Or there may be failure. It may be some personal failure such as a lapse of judgment that results in the follower's dismissal from the company, or the marriage might fail and the church turn its back on her, or the charismatic candidate may fail to win office and his movement falls apart. Then the follower faces different choices. Perhaps he may attempt to remain and try again, finding some rationale for the failure and investing the leader with new meaning to justify continued support. Perhaps he or she may seek another leader with a similar program. Or he may just leave in disillusionment, despising his gullibility and resenting the hopes aroused in him by the leader. There is a strong temptation to blame the leader for whatever failure has occurred, and indeed he may well have behaved poorly, but the follower may also succumb to blaming the leader for his own failings. At the most extreme he may leave and become a 'career apostate', who develops a new great work by opposing the leader at every turn (Foster, 1984), using hate to defend against the love that he once felt for the leader, a love at once so tender and guileless (Moore, 2001).

Or perhaps, even though there has been some failure at the surface level of the stated agenda, the follower may nevertheless still feel that she has got what she came for at a deeper level; perhaps new skills, confidence or expertise, or even the successful negotiation of a difficult life passage (Levine, 1984). It is disappointing that the charismatic world-changing agenda was not realised, and that the leader turned out not to be the person he was thought to be, but the follower has learned much and may now feel ready to move on and tackle other challenges.

The Evolving Relationship

What colours the success or failure of the great work is the evolving relationship with the leader. This relationship arcs from enthusiasm to abandonment; recall how British voters rejected Churchill at war's end, and how Freud dropped Fliess after he had gained what he needed. There may be stages along the way, minidisillusionments, renewals, crises of faith, disappointments and conflicts, but because a higher purpose is involved, both leader and follower may choose to endure these disruptions and to continue their relationship (perhaps in modified form) in pursuit of their higher goals. Of course, there must be some basic fit between their purposes; a Christian leader will not modify his religious beliefs to appeal to unbelievers, and the followers will only select a leader whose values and goals reflect their own. But the leader may temper his message to heighten his appeal, and the follower may ignore any unpalatable aspects of the leader's ideology. At that point they both hear what they want to hear, and say what each thinks is necessary for the relationship to progress. But most of all, the relationship follows the follower's trajectory rather than the leader's, because the follower is progressing emotionally whereas the leader is regressing. The leader will never outgrow his need for the follower to reflect his narcissism, whereas the follower, if he continues to develop, will definitely outgrow his need for the leader's guidance.

No matter how much commitment and love was present initially, through involvement over time the follower grows and learns and comes to understand the leader differently, perhaps seeing through some of his vanities and shortcomings. This changed perception is sharpened as the follower achieves his great work and becomes less reliant on the leader. There is invariably a reassessment of the leader and a scaling down of the idealisation and love that was at first felt. There is often disappointment; only a very few followers continue their associations with the leader for the rest of their lives in the manner that they began. The choice really concerns how the relationship ends, whether on a sweet or bitter note, for end it almost always does. At best the follower may simply move away quietly with his gains; at worst he may renounce his faith in the goodness of the world, dismiss his trust as naiveté, lose his courage for risk-taking and project his own darker side onto the leader, seeing him as an embodiment of evil.

The leader also becomes disappointed with the follower, although he will not usually end the relationship because of this. The leader's hopes for the follower founder on the follower's growing autonomy, which the leader experiences as a threat. The follower's personal growth and autonomy pull the leader back into realistic relations and contradict his need to feel special. The leader had hoped that through the follower he would gain the unqualified love and elevation that he 'knows' is his entitlement, and at first he did. But as the follower grows emotionally and achieves her great work she becomes less subservient, and may even come to expect equality, a reasonable expectation from her point of view but an affront from his. Eventually, the leader comes to see the follower as ungrateful, and possibly even a threat, and she is subtly shifted aside. The high hopes that the leader had for her, that she would develop into an ever more ego-flattering mirror of his narcissism, are disappointed by her inability to appreciate him on his own terms and her selfish and foolish desire (from his perspective) to follow her own path in life. He may mourn her limitations.

There are gradients and variants to this pattern. For example, many followers never actually meet or even go near their leaders. This may be simply a matter of circumstance, as with large-scale political leaders such as Gandhi, or maybe the follower does not want, or perhaps could not stand, to be so close to the object of their affections. In such cases they relate to an image of the leader that reflects their needs, creating a fantasy relationship with the leader in their minds. But, depending on the nature of the task undertaken, they may still go through many of the processes and changes involved in a more proximal involvement. One explanation for how this happens is that they hype themselves into it; another might invoke the placebo effect.

While a relationship exists between two people, its meaning exists within each person and may be different for both. For the follower it may be a relationship with oneself through the symbolic mediation of the leader who, whatever else they do, primarily functions as a catalyst. If the follower is creative he may come to feel the same quality of relationship with a leader who is distant, that another follower might feel with a leader who is close, perhaps as religious believers say they experience a living relationship with a saviour who died 2,000 years ago. In such a relationship there may be soul-searching, personal growth, crises, vanities discovered and fears dispelled, needs awakened and illusions explored. One may reflect, typically through prayer or meditation but also in dialogue with other followers, on one's behaviour and on the meaning of one's life and relationships. One may come to learn much about oneself. Such a relationship is one-way of course, but it is felt as being two-way. Yet even in such distant charismatic relationships, the follower may still come to outgrow their need for the leader.

In more interactive charismatic relationships the level of stimulation is much higher. Such relationships are more fertile and intense, but this may not suit every purpose. The British did not want to search their souls in their relationships with Churchill, they just wanted military victory. Nevertheless, membership in a charismatic movement inspires reflection and self-exploration (appropriate to the follower's awareness). The astute follower learns much about his or her self and, with increasing awareness, will glimpse the leader's limitations.

For there are times when the work has been done and the meal is served, and the follower seeks time out from their great work and the grandiloquence of charisma. Now the follower just wants to sit and commune. He may hope to get to know the leader better, to inquire about his personal side. When this happens he may discover that the leader really has no personal side, there is just more of the same. At first it merely seems odd that the leader cannot step out of their leadership role, or is only able to do so for short periods and cannot relate simply as an equal. While the leader may appear to do so, perhaps telling personal anecdotes and making jokes, there is that lingering hint of grandiosity and dominance just below the surface, and a suspicion that even at a personal level, everything that the leader says is said for an effect. Now the leader seems strangely rigid and repetitive, and the follower glimpses the price the leader has paid for his success; an inability to allow the 'free intercourse of unconscious parts of the mind — free association' (Mollon, 1993, p. 109). As long as there are no significant problems arising from this, the follower will probably ignore it. But problems do arise.

The Intimidating Effect

Perhaps the most significant change that occurs, even with followers who succeed in their great work and choose to remain with the leader, is the 'Intimidating Effect', in which the love felt by the follower gradually becomes tinged with fear (van der Braak, 2003). Quite early on in the relationship, the follower notices certain behaviours of the leader that inspire not love but anguish. Maybe the leader picks up on small points in conversation to expound upon at the follower's expense, or to impress or 'correct' him, or just because the leader needs to dominate. The leader cannot help doing this because his self-esteem depends upon raising himself above others, often by putting them down. He is most comfortable when others are off-balance; then he can function as a stabiliser for them, so from time to time he may deliberately act contrarily or arbitrarily. He may even perform some action that hurts others and belies his claim to legitimacy, and that he acts solely for their welfare; an action that demonstrates that ultimately he is in it for himself. Or, he may chasten one or more of the followers in the manner of an animal trainer showing who is boss; at such times the follower may detect an underlying threat in the leader's behaviour. Or, perhaps the follower sees the leader perceptively diagnose others' faults, and he worries what awful things the leader might see in him next, things he has not seen in himself. The follower may witness the leader unleash his anger on another who has fallen from favour, and he fears having the same thing done to him. The leader is freer in expressing his opinions than are most people and more confident of himself and, when in conflict, he invariably goes for the jugular; it can be hard to defend oneself against his unpredictable narcissistic rage (Kohut, 1972). Because of his experience, motivation and wide reading, the leader usually is more generally knowledgeable and better informed than others. He is invariably an agile conversationalist, and an astute critic who adroitly divides the world into good and bad; the bad being those who differ with him. If opposed, he doubles his energy to win this or that point. If further opposed, he doubles it again.

But most commonly there is just an accumulating weight of demands and criticisms that erode the follower's confidence in himself and alienate him from the leader. The inevitable result is that the follower learns that it is pointless to argue with the leader, that he must either accept the leader's dominance or leave. The follower may choose to remain and to keep silent about some things, but a tone is set in which the follower comes to see that even small points of difference risk becoming major combat zones in which the leader is ever ready to unleash the full power of his considerable intellect and aggression to win. This is especially so if the leader can count upon the support of the great majority of other followers who are still only part way through their great works, and thus are still dependent upon him and unwilling to oppose him. It becomes easier to let him get his way than to endure conflict, so dissent is suppressed.

This is corrosive of all that is healthy in the relationship, for it narrows it to a few issues that the leader feels comfortable recycling in order to remain dominant. There always remains much work and change ahead for the follower to undertake at the leader's behest. The corrosion occurs slowly and subtly, and may be marked by protests and challenges from the follower, but eventually a point is reached where the follower disagrees with much that the leader stands for, yet is intimidated into silence. The follower feels increasingly unseen and disempowered, and shocked to discover just how blind the leader can be, quite unlike how he experienced him at first. The leader is unaware of all this, being buoyed by grandiosity and with time the follower may even become afraid to leave because he knows it will precipitate a major confrontation and rejection. He may hang on in hope of achieving his great work, but if the tension becomes too great he will abandon the relationship in disillusionment.

There is resentment here, and for some followers their fear is actually a defence against their own feelings of murderous rage at having been dominated for so long. This rage can easily be activated should the leader fail or show weakness (for an example of this, see the case study of Savonarola in chapter 10). The recognition by the follower of his strong resentment may lead to guilt; he has gained much through following, so how can he feel such hate? He may even recognise that he is the one who has changed, and that the leader remains the same as he always was; so the follower wonders if he has any right to change the rules just because he now wants a different quality in the relationship. This leads to such confusion that, unable to distinguish love from hate and fear from need, the follower becomes unable either to remain or defect. Numbness may set in, but eventually the follower becomes desperate to leave because he feels beset with too many conflicting thoughts and emotions. This desperation at least enables the follower to leave by honestly saying that he does not know why he wishes to leave, he only knows that he 'has to get away'.

Another pathway arises if the agenda is changed. Typically, the leader motivates the followers with inspiring talk about the glorious goal to come, and with promises such as, 'When we achieve such-and-such we will be safe and happy'. But when the goal is reached the leader may lay out a new plan for struggle, hard work and sacrifice, in order to remain needed and dominant. From this the leader comes to be seen as arbitrary and inconsistent, and the follower guesses that these utopian programs will never end and that their real purpose is to perpetuate the leader in his leadership role. Even a highly motivated follower may baulk at such continual shifting of the goalposts.

Eventually, the astute follower glimpses an invasiveness to the leader, as if nothing short of complete dissolution of his being into the will of the leader will satisfy, and that the leader desires to possess him from within in some way, to absorb his identity into his own. The very things about the leader that the follower loved at the beginning — his strength, his vision and his energy — he comes to fear in the long run. For the leader actually demands much more than mere love — that was given at the start — but for the leader it was only the beginning. As Jerrold Post has written of Bill Clinton, at base Clinton 'does not want to be liked so much as be validated. He wants others to accept the view of himself that he holds, and when they don't he disowns them and turns against them angrily' (Post, 2005, p. 292).

Superficially the charismatic relationship may break down for all the same reasons that other relationships end, a quarrel or a misunderstanding or whatever, but the underlying reason is that the follower outgrows her need for the leader, and is no longer willing to tolerate the leader's unremitting, regressive need for dominance.

The Charismatic Crucible

This leads to the 'Charismatic Crucible', the struggle of the follower with his or her self. An excellent book by Andre van der Braak titled *Enlightenment Blues: My Years With An American Guru*, gives an account of this tortuous struggle by a devotee in a spiritual movement (van der Braak, 2003). The great strength of van der Braak's work is the honesty and detail of his self-observation concerning both his initial love for his guru, and his evolving process of disillusionment. He describes his many struggles with doubt and his battles with his conscience and his ego, his continual self-interrogation and rumination about intellect versus emotion, freedom versus commitment, the rights of self versus obligations to others, his attempts to strike a balance between faith and reason, his debates with other followers and with outside critics, while trying to sift the gold from the dross in his relationship with his guru. This is the Charismatic Crucible; it powers personal growth and is freely chosen, even warmly embraced by the followers.

Charismatic relationships invite such self-exploration, and it is in this examination that the potential for personal growth and change exists. The process is akin to psychotherapy, but is lived through rather than merely reflected upon. The Charismatic Crucible may inform the follower about him or her self; of the limits of their love and acceptance; of their ability to love, trust and forgive; of their adaptability and openness to experience; and of their capacity to endure heartbreak and the loss of cherished illusions. It may reveal the follower's true feelings about self and other; how honest they can be, how comfortable they are with their self and whether this is justified, and their capacity for both attachment and separation. There is invariably a testing of oneself against an absolute standard; as Rollo May wrote of his charismatic mentor Paul Tillich, 'Once in a great while it may be a good thing to have someone expect perfection of you' (May, 1973, p. 22). With this may come the revelation of both how big and how small one can be, for just as the relationship ultimately rests upon inspiration, much that is painful may flow from it; soul-food for a lifetime. As a man in *Prophetic Charisma* reported, 'My time with [his leader] was the most painful period of my life, but I wouldn't have missed it for the world' (Oakes, 1986, p. 10). It is a time of struggle with ultimate concerns (Barnes, 1978), of love, hope and freedom (Camic, 1980; Tucker, 1968), a time to examine the law (Sennett, 1975), to dream impossible dreams, and to suffer the consequences. But most of all it is a time to explore Love!

Through doubts and fears, highs and lows, strengths and weaknesses, the follower explores his commitment as he attempts to live by the ideals he has invested in the leader. Accumulating exposure to the leader brings gradual disillusionment as he comes to realise that the leader does not live by these ideals. As the quote that introduced chapter 8 expressed it, 'when it is no longer necessary to sort through the breviary of questions that concern his freedom' (Patton, 1980, p. xi), he leaves. The gains can be immense, but often they are resisted, for not every follower can tolerate such self-scrutiny. Of course, not all followers engage in such self-examination, the hard work of self-on-self when one frequently feels lost. Some run a mile at the merest whiff of it, though perhaps with regrets that they have passed up an opportunity to learn something special, and a suspicion that they may never feel so alive again, although these notions fade in time. Others only recognise the gains they have made belatedly, reluctantly and at a distance after leaving. An informant for the study that became *Prophetic Charisma* put it this way:

> After I left [the leader] I needed a job quickly so I took one with Metro Company. Now, they have a nasty reputation for the way they treat staff and I was a bit worried at first, but I found that nothing they did could faze me. I'd been fucked-over by an absolute master, and compared to that, these guys were just pussycats.

The Charismatic Moment

The charismatic relationship is not all struggle; if it was people would not enter it. What gives the relationship its unique appeal is recurring experiences of intense love, joy and meaning, which inspire the followers to greater communion, commitment and energy. Such experiences constitute what Charles Lindholm has called the 'charismatic moment' (Lindholm, 1990, p. 189). It seems that the charismatic relationship exists at four levels, or at least, that four purposive themes recur time and again. The first involves the stated purpose of the group, the leader's program of salvation or reform or whatever, and the recruitment by him of others to achieve this purpose. The second is the follower's great work, their personal motivation concerning what they hope to use the relationship for, what they personally hope to achieve, and their purpose for belonging. The third level is the Charismatic Crucible, the self-examination of the follower that is provoked by the relationship. This involves the soul-searching and interrogation of self and others that takes place within and between each follower (to the degree that they are willing and able to engage in it). The fourth and deepest purpose of charisma is to experience the powerful love that fuels the other levels. This experience of love is felt by the follower as an encounter with his or her ultimate concerns, in religious parlance: God (where God is love).

Several scholars have described, and some have elegantly measured, charismatic love. The experience is extraordinary and powerful.

In ecstatic moments the participants forego the rhetoric and agenda of the group, and bask together in rapture. It is akin to what Herman Schmalenbach (1961) described as 'communion', and which Victor Turner (1969, 1974) called 'liminal' experiences. Turner showed that in ritual processes, during pilgrimages, and in crisis situations such as bomb shelters, there can occur a special kind of relating in which individuals transcend role-bound behaviour. When people are in this mode they relate more from their essences in a spirit of love, than from their statuses in an exercise of power. Such experiences have been best researched in charismatic communal movements.

Benjamin Zablocki, in his 1980 investigation of American communes, studied networks of dyadic love. He concluded that 'undoubtedly the most significant' finding was what he called the 'love density effect' (Zablocki, 1980, p. 355). It turned out that the greater the amount of love present in a communal group, the less stable that commune was, unless it had a resident charismatic leader to enforce stability. So reliable was this effect that Zablocki was able to deduce from it a 'social thermometer' by which he could measure the degree of love in a commune, and from that predict what proportion of its membership would still be living there one year later, and even whether or not the commune itself would survive.

Raymond Bradley, a student of Zablocki, divided communes into four basic types depending upon the amount of charisma present. He then correlated the love and power relations that existed in each group, with the amount of charisma present. He found that the greater the charisma within a particular group, the more intense were both the loving and power relations inside that group. Bradley concluded that great power was needed to contain the inherently destabilising effect of love. He went on to suggest that love and power might be in some way related at some deeper level, as two sides of the same coin, in a way that optimised, enriched and stabilised social life. (He subtitled his book, '*A Study of Love and Power, Wholeness and Transformation*', Bradley, 1987).

These investigations, along with Lindholm's (1990) study of Adolf Hitler, Charles Manson and Jim Jones, suggest that beneath the multitude of rationales that members of charismatic groups give in order to explain their reasons for joining, there lies another, perhaps unconsciously held agenda: to experience the intense love that heals and transforms. In charismatic groups, love may be experienced so powerfully that it threatens the social fabric and the very existence of the group, and thus must be contained by strong bonds of power if the group is to survive, though power corrupts and hence even at their best, charismatic groups are unstable. This makes the relationship between love and charisma purposive; the charismatic leader presents himself as a catalyst through which others explore their loving. To do this he devises rituals that provide opportunities for this love to be felt. These rituals may be as modest as church services or as vast as a Nuremberg Rally. At least for some they are transformational, and the leader may also hope for healing or transformation. But it takes courage and persistence, and many fail.

Endnotes

1 This is a genuine instance when individual psychology and group psychology become mirror images of each other. The leader acts out a role that the followers would love to enact (at least, their versions of it), and in the followers' mirroring of him an image of the group is activated that spreads to influence all it touches, and within which each player plays their role, while also carrying within their heart the hearts of others, and being carried and sustained by them.

2 See Bass & Riggio (2006, p. 4); Brown & Moshavi (2002); Brown & Trevino (2003); Kahai, Sosik, & Avolio (2003); Zohar (2002).

3 See Dumdum, Lowe, & Avolio (2002); Lowe, Kroeck, & Sivasubramianiam, (1996); McColl-Kennedy & Anderson (2002); Vandenberghe, Stordeur, & D'hoore (2002).

• • • • •

CHAPTER TEN

· · · · ·

The Charismatic Movement

Nothing outside his own self held any significance for him, because everything in the world, it seemed to him, depended on his will alone.

Leo Tolstoy

After the defeat of Germany in 1918, Adolf Hitler suffered his emotional collapse, then recovered and continued to search for his place in the world. He had little success at first, and he stayed afloat by working as a government informer. This took him into the political underground, a world of fringe parties with extreme ideas, of street fighting and pamphleteering, of strange groups such as the Thule Society, a quasi-occult, anti-Semitic, semipolitical crank group that included several senior Nazis, and of conspiracy theories, state secrets, betrayals, assassinations and bomb plots. As a double Iron Cross holder he had great prestige in this milieu. His exceptional intelligence gave him leverage, and his artistic temperament gave him unusual insight. He cut quite a different figure from the confused, lost youth he had been, and he was psychologically distinct from the angry ex-soldiers and mainstream aspiring politicos he now mixed with. Soon he began to rise, joining the fledgling Nazi party and then becoming its spokesman. His discovery of his talent for public speaking was the breakthrough that gave him confidence and established him as a leader, while also cementing his self-concept as a visionary.

Hitler's discovery of his public speaking talent transformed him. Soon he was claiming that the great ideas of history first emerged in the spoken word rather than as written text, a convenient conceit that elevated him to the status of a prophet. Given his earlier premonitions of his destiny, this was not inaccurate. His speaking ability, which had started out as a self-soothing technique that eased the pain of his abusive childhood and his failure to fit in with his peers, became elevated in his mind as another indicator of his special calling. It was this ability to voice his vengeance at the world in sibylline rhetoric that had so inspired his young friend August Kubizek when they were alone together beneath the stars. As Kubizek recalled, it was as if 'another being spoke out of his body, and moved him as much as it moved me ...', and that Hitler too had 'listened with astonishment and emotion to what burst forth from him with elementary force ... a state of complete ecstasy and rapture ... of a special mission that one day would be entrusted to him' (Waite, 1977, pp. 213–214).

What young Adolf had achieved was the trance state that mystics, mediums, shamans and firebrand preachers are renowned for, and when the time was ripe others responded as Kubizek had done. In Hitler's usage, oratory became a powerful emotional ritual of intense arousal that moved people to tears. At the Nazi rallies, captured in Leni Reifenstahl's 1938 film *Triumph of the Will*, men shouted 'Sieg heil' and women wept in a passionate mass ritual of oneness and ecstasy. Joseph Goebbels kept a diary where he described attending party meetings where Hitler spoke, and of his burgeoning enthusiasm for him, writing that Hitler, 'has got everything to be a king ... a genius ... Adolf Hitler, I love you' (Heiber, 1962, p. 48). So powerful were these ritual unions that even many years later during the Nuremberg Trials, after von Ribbentrop was shown a film of Hitler, he burst into tears and exclaimed, 'Can't you see how he swept people off their feet? Do you know, even with all I now know, if Hitler should come to me in this cell now and say 'Do this!' — I would still do it' (Gilbert, 1950, pp. 195–196).

Max Weber dubbed such intense experiences 'pure charisma'.[1] This involves an immediate and undiluted, face-to-face experience of love, oneness and excitement, in which one identifies so totally with the leader as to come to see him or her as the source of one's own ultimate good. Weber outlined a continuum ranging from 'pure' to 'routinised' charisma. Pure charisma is rare and is usually found only at the beginning of a social movement when a 'charismatic community' coalesces around a leader. This community is characterised by a belief in the special talents of the leader, intense emotional bonding by the community to him and estrangement from the world as a whole. Thus pure charisma is personal, and is based on face-to-face contact and feelings of trust, duty, and love on the part of the followers. It is creative and revolutionary for, 'in its pure form charisma ... may be said to exist only in the process of originating' (Weber, 1964, p. 364).

At the other end of the continuum, routinised charisma occurs when the leader's charisma is thinly dispersed throughout the followers who act in his or her name, typically after the leader has died. It may survive many generations and underlie a stable social order, but it is conservative and is no longer a force for social change. Hence, in Weber's formulation, while charisma at its purest is face-to-face, passionate, loving and creative, at its weakest it may be little more than a claim by some official to act in the name of a dead or absent charismatic leader.

Weber used both religious and economic factors in his social theory. He saw Western civilisation as moving towards greater and greater rationalisation of all aspects of life. This, he feared, would make life into an 'iron cage' and turn daily existence into an alienated, mechanical, meaningless routine. But Weber also believed that ideas, especially religious ideas, can profoundly influence society, and that they cannot be simply dismissed as a function of underlying social, cultural, economic or psychological processes. One source of new ideas involves the periodic emergence of charismatic personalities.

As is clear from Weber's original definition, charisma is not solely a quality that a person actually possesses. It is dependent upon the perceptions of the leader by his followers, those who see the leader as divine, for they are as much a source of his power as are his personal talents and without them he is nothing (Weber, 1968a, pp. 241–242). Weber focused especially on prophetic charisma that occurs in complex societies, rather than in tribes, and which adheres to the prophet who proclaims a divine or radical doctrine, typically a Jesus or Buddha figure. This form of charisma leads to revolution and social change, and the charismatic prophet claims authority by sheer force of personality. Weber regarded the prophet as the prototype for other kinds of charismatic leaders. The prophet points to some mission outside or beyond himself that he embodies, and his mission involves the radical change of current values. Before receiving his calling such a leader must have some germ of charisma latent in him, but later he maintains power solely by proving his strength in life; to be a prophet he must perform miracles (Weber, 1946, pp. 248–249).

The prophet arises in a time of crisis, as the Old Testament stories attest, typically when someone from outside the powerful elite and the administrative bureaucracy presents a critique of what is wrong with society and how to fix it. Claiming to be a messenger from God, or sometimes even God incarnate, the prophet opposes — even if only with words — the status quo, and demands change. Either there is to be a return to the virtuous ways of the past, as with, 'Repent for the kingdom of God is at hand', or there is to be an embrace of some new revelation, as with, 'It is written, but I say unto you ...'.

Weber wondered whether charisma might arise from some mental illness. He rejected this notion and instead spoke of an emotional seizure that originates in the unconscious of the leader and results in extraordinary emotions. Weber placed these emotions on a continuum ranging from a high point of ecstasy to a low point of passion, or mere enthusiasm. To incite enthusiastic political passions, the charismatic political leader uses rhetoric and mass gatherings such as political rallies. To arouse euphoria, the charismatic evangelist uses ritual, song, and readings from sacred texts. To produce ecstasy, the shaman or prophet may use drugs, music, dance, sexuality or some combination of these; 'in short, orgies' as Weber wrote (Weber, 1968b, p. 273). The shaman or prophet may also provoke hysterical seizures that can seem like a mental disturbance or spirit possession. These emotions serve to recruit and motivate followers and the greater the leader's emotional depth and belief in his calling, the greater is his appeal and the more intense is his following.

Thus Weber saw charisma as essentially a religious concept that powerfully influenced and penetrated secular society. Unfortunately, both the technical sociological usage and the popular corruption of the term tend to overlook this religious sense. It was left to sociologist Thomas Dow to restore the original sentiment of the idea by returning to Greek mythology, to the original meanings associated with charisma (Dow, 1978). He showed that in essence charisma represents a daemonic creative force, 'the thrust of the sap in the tree and the blood in the veins … the incarnate life force itself' (Arrowsmith, 1958). Charisma 'revolutionises men from within', is elemental rather than ethical — the reality of careless power — and in its most potent forms it is subversive of all notions of sanctity. Invoking Neitzsche's terminology of Apollonian restraint and Dionysian release, Dow argued that charisma is essentially Dionysian, a state of being beyond reason and restraint, and of freedom from social controls. Charisma rejects all external order, transforms all values and compels 'the surrender of the faithful to the extraordinary and the unheard of, to what is alien to all regulation and tradition, and therefore [it] is viewed as divine' (Weber, 1968a, pp. 1115–1117). The subjective state of charisma is ecstatic freedom.

Charisma exists when an individual's claim to exceptional gifts is acknowledged by others as a valid basis for their participation in an extraordinary program of action that takes them beyond everyday routine, custom, law and tradition, and that lifts them out of themselves (or perhaps evokes their deepest selves). The charismatic leader is viewed by the followers as a model of release, and of the divine power that makes freedom possible. The follower recognises in the leader forces that exist in him or herself. The courage the follower needs to abandon his self, and to overcome the limits of internal and external controls, is provided by identification with the leader. But the followers surrender not so much to the person of the leader as to the power manifest in him, and they will desert him if his power fails. The followers attain freedom by surrendering to the leader, and through him, to their own emotional depths. This is their Good, not in some ethical or conventional or intellectual sense, but in a primordial or instinctual way. Ecstasy comes from breaking down inhibitions, from the experience of carefree power and from the abandonment of conventional morality. Charisma is thus an emotional and spiritual life force opposed to law, conformity, repression and the dreariness and predictability of orderly life.

What might such radical freedom look like? There is a brief vignette by Henry Suso, a 14th century Christian mystic, about a conversation between himself and a spirit visitor that describes the heart of charismatic release.

> Whence have you come? (Suso asks).
>
> I come from nowhere, (the spirit replies).
>
> Tell me, what are you?
>
> I am not.
>
> What do you wish?
>
> I do not wish.
>
> This is a miracle. Tell me, what is your name?
>
> I am called 'Nameless Wildness'.
>
> Where does your insight lead to?
>
> To untrammelled freedom.
>
> Tell me, what do you call untrammelled freedom?
>
> When a man lives according to all his caprices without distinguishing between God and himself, and without looking before or after.[2]

Weber's definition of charisma has two parts. First (unlike St Paul), not just anyone can be charismatic, at least in this leadership sense that Weber focuses on. Second, charisma implies a relationship; it has to be recognised by others in order to exist. No matter how successful or clever one is, unless one inspires hope, love and devotion in the hearts of others then one cannot be considered charismatic. This results in a slippery definition that relies upon reference to both inner states and recognition by others. The slipperiness could probably have been avoided if Weber had argued that only persons with a particular disposition could become charismatic, and that most of these would never do so because their life situations were unfavourable, thus allowing for social influence. That is what we find in the real world; many are called but few are chosen. Some are burdened by mental disturbance. Others lack the crucial interpersonal skills to recruit and manage a following. Many get passed over by history, never quite capturing the hearts of those who might become their followers, or never finding themselves in control of sufficient power to realize their vision. A few eschew leadership.

Nevertheless to Weber, and to the ancient Greeks and to Paul, charisma is basically a religious concept. It shows the spiritual interacting with the secular. The charismatic leader's mission is based on spiritual authority, pointing to some great purpose or ultimate concern beyond whatever current program in which the group is engaged, and signifying release from stale social restraints and new life in some form. Despite Weber's focus on very large-scale charisma, it may also be found among the low and the high, in office politics and international affairs, in the academy and the media. Charisma's unique appeal derives from the recognition by the followers of a correspondence between their own needs and the personality and program of the leader. They follow not for mere material gain but in the hope of something like salvation, or personal transformation, or from utopian yearning. In following, they experience an emotional freedom that revolutionises them from within, as Hugh Milne, a former follower of Bhagwan Shree Rajneesh, has described:

> Many people have asked me how a sensible, independent person could be mesmerised by someone like Bhagwan. The answer, as many sanyasis would agree, is that once you had been affected by his energy and experienced the sensation of being touched by it, you knew there was nothing like it, no bliss to compare with it. Once you had experienced it you had to go back for more, to try and regain that feeling of harmony and being at one with the universe. It is similar to a drug induced high except that there is no chemical at work. Bhagwan's touch could be just as addictive as the strongest drug. (Milne, 1986, p. 179)

Charismatic movements have long fascinated observers, who also ask such 'how' and 'why' questions as Milne alludes to. Two of the best studies of such movements were those done into American communes by Benjamin Zablocki and his student Raymond Bradley (Bradley, 1987; Zablocki, 1980). Although very different from such large-scale movements as the Nazis, the very intimacy of such groups and their relatively modest scale and stability, render them better subjects for study. As discussed in chapter 9, beneath the multitude of rationales that members of charismatic movements give for their involvement there lies a deeper agenda: to experience the intense love that heals and transforms. Thus the relationship between love and charisma is purposive; members join charismatic leaders not only for the agendas espoused, nor just for their own advancement, nor solely for the self-exploration that is stimulated in the Charismatic Crucible. As much as all of these, they are attracted by the experience of love that is stimulated by the leader, as the Goebbels quote (above) suggests.

The charismatic movement is an invitation to join in a great utopian effort, and an opportunity for the followers to further some Great Work of their own. But in order to keep the followers motivated, the charismatic leader needs to provide charismatic rituals of transcendence wherein they can experience the love and fusion that they seek. Whether this is done through mass rallies as with the Nazis, or intense prayer meetings as with Pentecostal churches, or just motivational seminars for employees, the basic pattern is the same. Members are provided with novel and exciting stimuli that enable them to rise above themselves, to cast off inhibition and restraint, and to open up to and join with their comrades in a deeper way than they usually do; to abandon roles and to enter a liminal state of deep communion with each other and to experience themselves in the extraordinary ways that Victor Turner described (Turner, 1969, 1974). From such ritual experiences they emerge refreshed and inspired, reinvigorated in their commitment to the leader, to each other, and to the Great Work that each is attempting to perform. As Charles Lindholm has argued, the justification and function of this is that society is not based on rational processes, but upon a deeply evocative communion of self and other, a communion that offers not reason but lived vitality, and that without this electrifying blurring of boundaries life loses its savour, action is no longer potent, and the world becomes colourless and drab (Lindholm, 1990, p. 189).

All societies permit such experiences that echo the shamanic rite, and they may be as trivial as football matches (Cole, 1973) or as profound as the Catholic Mass. They restate core values and integrate the participants into a communal bond. Often they make use of natural life transitions such as coming of age, birth and death. What is distinctive about charismatic rituals is that they are focused upon an individual, the leader, who is always near the centre of the action. They are also much more emotional and intense, for the leader aims to realise a new revelation, or perhaps to reinstate an earlier revelation, rather than merely to reinforce the status quo.

Such rituals are the leader's main creative accomplishments. He permits a greater access to himself than the followers usually experience, answers many of their questions, and provides an example through his own behaviour and/or teachings of what the exalted condition they are seeking is actually like. His very proximity is arousing. He seems to demonstrate whatever is 'just right', to speak to each one individually, to read their concerns before they themselves voice them, to reveal them to themselves and to draw them out of themselves into a new life beyond their fears and egos, to reveal 'the beyond in the midst of life' in Dietrich Bonhoeffer's happy phrase.

However, there are conditions upon such charismatic love that neither Zablocki nor Bradley mention. First, there is an underlying principle that limits it to the movement. If one attempts to move out of the movement, the love will be withdrawn. Similarly, if one queries or challenges the leader it will also be withdrawn. Charismatic groups thrive on enthusiasm and hope; to dissent or defect is to disturb the happy flow of communion and to raise doubts. It is as if there is a principle that runs: the success of all depends upon the agreement of all (or at least upon agreement on the core assumptions about the leader and the group's stated agenda). This qualifies the love that is generated.

A charismatic movement is also a mosaic. By definition it is a minority. But only a minority of this minority are really there for idealistic or transformational reasons, and only a minority of that minority actually realise their hopes. Some join for purely opportunistic reasons, and they are not forced to develop themselves. Some manage to live around a charismatic personality for years and to not grow or learn much at all. When considering charismatic groups one must be aware that there is often much more noise than signal. Very few of the followers remain devoted all their lives, and there are the inevitable lame ducks, misfits and eccentrics who tag along. Some of the followers may have strong personalities, and in other contexts might themselves be leaders, but they fall short of the required leadership qualities in that context because they are damaged in various ways. They remain committed, partly in order to realise their own hopes and partly because they become emotionally dependent. In sum, the meanings of the charismatic rituals and the experience of love vary from follower to follower.

Much depends on circumstance. A charismatic revolutionary like Fidel Castro may seize power with the help of many strong, intelligent, mentally healthy followers, and he may retain their loyalty because there is much important work to be done. The followers may even become quite pragmatic about him in the course of the years; it is not impossible for charismatic relationships to evolve into realistic relationships, though the chance is small. Are the ones who remain as followers merely dependent losers? Some perhaps, but most are not. One may feel obliged to remain for the sake of one's children or spouse, or for other equally realistic reasons. But only a very few followers experience such a comfortable 'fit' with the leader that they are happy to remain enthusiastic followers for many years. Charismatic movements lose far more members than they retain.

The charismatic movement is really quite a remarkable achievement, for along with the followers who sincerely believe in the leader, most such movements also include followers who do not believe, or at least not in any very strong way, and some who do not really believe in anything much at all. There may even be a few who actively oppose the leader. As well, even among the followers who are sincerely enthusiastic there are conflicts and personal differences, with some following for what might appear to others to be all the wrong reasons. When one considers the sheer perversity and idiosyncrasy of human beings, and then remembers that it is love, not force, that bonds them, the result seems astonishing indeed. Perhaps only a charismatic leader, or some great threat, could forge them into such a functional unit.

However, there is a downside, for the charismatic movement and charismatic leadership generally is highly unstable. There are several reasons. First, power corrupts; even at his most successful and effective the leader faces daily temptation to indulge in selfish or antisocial behaviour. Second, charisma lacks restraint, and without the restraining hands of others the leader may succumb to unwise actions or decisions. Third, charisma lacks institutional support; its only legitimacy is its continuing success and when it fails there is no historical or rational reason to support it. This last point also reveals that charismatic leadership is a constantly contested niche, subject to threats from dissidents and outsiders. Because of the lack of institutional supports there is usually little or no accepted procedure for the transfer of authority; this problem of succession infected even a major religion like Islam. The most unstable form of charismatic leadership is prophetic charisma (Oakes, 1997), wherein the charismatic leader identifies 'the millennial destiny of mankind with their own personal vicissitudes, and demonizes any opposition to their aspirations and personal aggrandizement' (Robbins & Anthony, 1995, p. 244).

In the face of instability, corruption, or some failure of charisma, and lacking institutional supports or restraints, the charismatic movement may become increasingly unhinged, violent or extreme, perhaps even turning upon itself or the leader in hatred and/or disillusionment (Le Bon, 2006). Sigmund Freud saw this very clearly, perhaps because he was surrounded by the irrationality of his own followers for many years, albeit at his behest. His son Anton has described how, 'The most trivial platitude swelled up in importance, and nobody seemed to treat Freud like a normal human being, many of the visitors playing the part of courtiers at what was both a psychoanalytical and a private court' (Behling 2005, p. 132). Freud's judgment was that, 'The group is extraordinarily credulous and open to influence, it has no critical faculty, and the improbable does not exist for it. It thinks in images' (Freud, 1921, p. 13).

This relates strongly to the theory of charismatic leadership advanced by Graham Little (1985). Little's theory was an attempt to explain contemporary political movements in terms of their underlying psychodynamics. He argued that because of their early life experiences, most people were predisposed towards one of three broad 'political ensembles'. The left-wing ensemble arose from early life experiences of need, and thus was attuned to social justice and helping. The right-wing ensemble arose from early life experiences of threat, and was oriented towards strength and structure. Charismatic leaders arose from narcissistic early life experiences, and their movements were oriented towards hope. According to Little, charismatic movements were small and were most likely (although not invariably) to combine with left-wing ensembles to achieve their agendas.

This left a hole in Little's theory because it implied that right-wing ensembles ought usually to be minorities, but this clearly is not the case. This problem is solved by postulating a fourth grouping (or quasi-ensemble), that is also small but which may combine with any ensemble (though usually a right-wing ensemble) to achieve its agenda. This can be described as a sociopathic grouping, arising from early life experiences of anger and oriented towards hate. Even in the most civilised of societies there exist many individuals who can be recruited by opportunistic leaders to antisocial ends. There are also many hapless conformists who, depending upon the context and without initiating any actions of their own, accede to quite horrific social agendas by averting their eyes. Add to these the 1% or 2% of the population who are clinically psychopathic and who are prepared to do the dirty work of any regime, and we have the latent risk factor in every democracy.[3]

Every charismatic leader attracts some such individuals, and in hard times the temptation by the leader to rely solely on these 'willing executioners' (as a famous book put it; Goldhagen, 1996), may be intense. Some leaders court such individuals; Hitler's remark that 'You can't fight a war with the Salvation Army' indicates his attitude; (he had just read General Blaskowitz's report of the Nazi genocide in Poland that described the Gestapo as 'subhumans who do not deserve the name German'; Baker, 2008, pp. 156–157). But even very prosocial charismatic leaders may find uses for such persons, and may become vulnerable to them during difficult periods. It seems that Freud and Gustave Le Bon and other extremely negative group theorists were actually describing a psychopathic ensemble, or something like it (for clearly, not all groups are as irrational, dangerous and emotional as Freud suggested).

Hence, one danger for charismatic movements, especially those operating from the right (the left has other problems), is that they recruit individuals whose sensibility is basically sociopathic, and whose hatred may become directed against the movement itself. An excellent historical example of this simmering, unstable irrationality of the charismatic movement, and how it can swing wildly and unpredictably out of control, concerns the 15th century movement known as the Weepers that was headed by the Catholic priest Girolamo Savonarola (Martines, 2007).

In the second half of the 15th century the West was undergoing dramatic changes. The Renaissance had arrived with its wealth and splendour, and the forces that would ultimately impel the Reformation were underway (Martin Luther lived from 1483 to 1546). Political, religious and economic tensions were high. Into the mounting storm came Savonarola, son of a wealthy Florentine family who, after being rejected by a young lady seems to have retreated into extreme religious moralism. He joined the Dominican Order without telling his family, and he took it upon himself to oppose the pagan excesses of the Renaissance.

At first his preaching was a dismal failure, but after being drawn to the Book of Revelation his sermonising took an apocalyptic turn that caught the discontent of the times, and he became a great success by preaching hellfire, repentance and damnation. All Florence thronged to hear him, falling to their knees in ecstasy and tears (hence the name, the Weepers). He made several prophecies that seemed to come true, specifically foretelling the death of Pope Innocent VIII, the arrival of a large foreign army as a scourge of God (this was the army of Charles VIII of France that occupied Florence for a time), and the collapse of the Medici family. These prophecies greatly increased his influence, which was described by Machiavelli:

> The people of Florence are far from considering themselves ignorant or benighted, and yet brother Girolamo Savonarola succeeded in persuading them that he held converse with God. I will not pretend to judge whether it

> was true or not, for we must speak with respect of so great
> a man, but I may well say that an immense number
> believed it without having seen any extraordinary manifes-
> tations that should have made them believe it.
> (Machiavelli, p. 149)

With his enthusiastic following came power, and with this power
Savonarola set about reforming the monasteries and greatly increasing
their populations. Next he turned his attention to the laity, denounc-
ing the Medicis, then rulers of Florence and leaders of the humanistic
revival in the city, and he even managed to drive out the hated Pietro
di Medici. He instituted the 'Bonfires of the Vanities', in which his
moral police went from house to house each Sunday calling out for
paintings, sculptures, fineries, luxurious clothing, books of poetry,
jewellery and ornaments — everything that he claimed drew men
away from God — to be given to them to be burned. Gambling was
forbidden and playing cards and dice were confiscated, yet still his
following grew, drawing as it did upon the anxieties of people experiencing a great social transition. By 1492 he virtually controlled
Florence, and he instituted a kind of theocratic democracy based on
his own interpretations of scripture.

Success only made Savonarola more extreme. After being called
to Rome to defend his actions — a call he disregarded — he
denounced the Pope and was excommunicated, which only made
him even more defiant and his preaching yet more passionate. He
eventually wrote letters to other rulers in Christendom urging them
to join him in opposition to the Pope, a not unreasonable hope due
to his continuing alliance with Charles VIII.

This was going too far and his enemies now moved against him.
On the order of Pope Alexander VI he was formally arrested for
investigation, and an adversary from the Franciscan Order agreed to
undergo the Ordeal by Fire to test whose teachings — Savonarola's or
those of the Church — were correct in the eyes of God. In this ordeal
both the Franciscan and Savonarola would be scourged by fire, with
the survivor being declared the bearer of truth.

At this point the mentality of the mob seems to have taken over.
Although Savonarola did not want to take up the challenge, some of

his enthusiastic adherents declared themselves ready for it, and word spread that he would indeed undertake it. When the fateful day arrived a large crowd gathered to witness the divine at work, expecting a miracle. Then word arrived that the Franciscan adversary had withdrawn his challenge. No doubt greatly relieved, Savonarola withdrew also, believing that it was no longer necessary for him to be tested. But by now the crowd, in fever pitch, demanded that Savonarola perform the miracle anyway to prove that he carried the word of God. At his refusal they turned upon him. He was seized and turned over to church authorities, tortured and forced to confess that his prophecies were not from God. He was declared a heretic, hanged and his body burned at the stake. Nevertheless his most loyal followers hailed him as a martyr and a saint.

The logic of Savonarola's appeal was simple; in uncertain times people fear the worst, and at that specific point in history, anyone who could argue for a connection between societal stress and the biblical apocalypse was sure to gain a following. By arguing for a return to moral purity as described in the Old Testament, Savonarola merely followed an ancient tradition. He presented himself as a prophet, and when his prophecies seemed to come true, people responded as if to the saviour. His lucky prophecies probably deepened his own conviction in the correctness of his mission. But Savonarola misread the context. He failed to appreciate that the real cause of the people's distress was the religious, economic and political turmoil of the times, and that the chain of transformative events that led from the Renaissance to the Enlightenment was far greater than any one man could direct. His conceit was to imagine that his power came from God, rather than the coincidence of his talents with the mood of the people. The most he could realistically have hoped to achieve was to serve as an example to others. Instead he tried to be a demi-god.

Savonarola was at first filled with zeal, piety and self-sacrifice, but as his success grew so did his extremism and obstinacy. He came to believe his own excesses, seeing himself as an apocalyptic prophet called to redeem all Christendom. But the crowd are unpredictable, superstitious and emotional, and any charisma based upon them is inherently unstable. Lacking the tools — a Gestapo or a KGB — to enforce his will, or at least to defend himself, Savonarola fell victim to the whim of the crowd, always desirous of a good spectacle. By taking the easy option of not undergoing the test (for the obvious reason that he knew he would be killed), he showed the crowd that they were, after all, led by a mere mortal and that he himself had no faith that God would stay the fire, or that God was leading him and would protect him. This dashed their hopes for transcendence, showed that their enthusiasm was, after all, just human emotion rather than divine passion, and that they were not participating in the end-time, merely following another zealot. They turned on him in outrage at what they felt was his trickery and betrayal, except for the few for whom no disconfirmation was sufficient to shatter their faith. To this day church factions still argue over his status as saint or charlatan.

Savonarola was really a victim of his own self-belief, for he seems to have come to believe that he really was a magical figure. He did not believe so totally as to risk the test, unlike Simon Magus who, according to Hippolytus, was buried alive at his own request, so confident was he that he would arise on the third day (he did not), or the 12th century Arab messiah who offered to have his head cut off to prove his authenticity and who also died (Wallis, 1943). Such leaders' hubris and their followers' enthusiasm may feed off each other, and magnify to psychotic proportions the leader's grandiosity. There was also rigidity and concreteness to Savonarola's beliefs, and an impatience and impulsivity to his actions that is striking, as if he literally believed in the absolute truth his calling, and this bound him in a particular direction with few alternative courses of action. Of course, charismatic personalities invariably have burgeoning confidence in themselves, it is one of their defining characteristics, but the savvy charismatic who can adopt a wait-and-see, one-day-at-a-time flexible tentativeness regarding their beliefs, has a distinct advantage over one who literally and fanatically believes in their own destiny. One of the limitations of contemporary Western spiritual leaders who adopt Eastern religious thought, typically Da Free John and Andrew Cohen, is that they take their religious systems quite literally; one who can place himself at one remove from — or above — his belief system, has additional freedom of movement, although such cutting oneself loose from all restraint may lead to other problems (Rawlinson, 1997).

Endnotes

1 Weber 1946, 1958, 1964, 1968a 1968b. There have been numerous criticisms of Weber's theory, perhaps most significantly by Bradley (1987), and there have been some modifications to some aspects of it (e.g., Berger, 1963; Sennett, 1975), but little change in his key concepts (Blau, 1963; S.C Olin, 1980).

2 Henry Suso, cited in Zerzan (1988, p. 14).

3 Some writings, such as the essay by Jonathan Haidt titled, 'What Makes People Vote Republican?' (Edge; The Third Culture, 9–9-2008; www.edge.org/3rd_culture/haidt08/haidt08_index.html), may be construed as attempting to legitimise this ensemble. Of course the members of Little's right-wing ensemble may legitimately hold conservative values, but those individuals who advocate or perform antisocial or racist behaviours, are more likely to represent the sociopathic ensemble than mainstream conservatism.

CHAPTER ELEVEN

• • • • •

The Charismatic Woman

[Women's] capacity for identification is not an expression of inner poverty, but of inner wealth.

Helene Deutsche

Why are there so few charismatic women? The handful one can easily name — Catherine the Great, Elizabeth I and Indira Gandhi — merely serve to demonstrate just how rare they really are (Bass & Riggio, 2006, p. 112). Yet in organisational settings women are often perceived as more transformational than men. Women tend to be better emotional communicators than men, to have more refined interpersonal skills, to be more relationship-oriented than men and to be more sensitive to the developmental needs of others (Bass & Riggio, 2006; Burns, 1978; DePaulo & Friedman, 1998; Eagly, 1991; Riggio, 1992). So pervasive is women's superiority regarding communication skills and interpersonal sensitivity, that management magazines have for some time been touting 'the feminine advantage' in leadership. Hence, a recent meta-analysis of 45 studies that examined sex differences in leadership concluded that women leaders were more charismatic than male leaders (Eagly, Johannesen-Schmidt, & van Eigen, 2003). However, it has also been noted that some traits favour men as charismatic leaders; specifically, men tend to be more self-confident, less conforming and more risk-taking than women (Bass, 1985; Hennig & Jardim, 1977). When men and women work together, men are quicker to claim leadership than women, and groups tend to look to men for leadership unless there is some situation of internal conflict, in which case women are the preferred and most effective leaders (van Vugt, 2008, p. 44). Scholars have interpreted that as indicating that 'a history of inter-group conflict might have predisposed men to adopt a hierarchical leadership style, while a need for social unity might have equipped women with a more egalitarian, personalised and communal style' (van Vugt, 2008, p. 44). Now, some researchers have concluded that charismatic leadership actually requires a balance of masculine and feminine traits (Hackman, Furniss, Hills, & Paterson, 1992).

The easy answer as to why there have been so few female charismatics is that there has been no social niche for them. In all cultures great power is invariably defined as masculine, and the symbolism and imagery of power pertains to men rather than women, although most cultures have their notable exceptions, Boudicca (Boadicea) for example. Even clinically, narcissism is thought to be more a masculine than a feminine trait (Rienzi, Forquera, & Hitchcock, 1995), perhaps because men are widely considered to be more aggressive and self-entitled than women (Tschanz, Morf, & Turner, 2008). However, this answer does not fully satisfy because charismatic personalities thrive on challenge; opposition and conflict bring out the best in them. A charismatic woman would be well able to overcome social limitations and transcend gender-based restrictions if she set her mind to it, perhaps in the manner of Joan of Arc.

But gender roles do count for much, and the charismatic woman may have to make sacrifices. Some, like Madame Blavatsky, founder of the Theosophical Society, stay unmarried and childless. In fact, so limiting can gender roles be that charismatic women often recruit a male lieutenant to deal with routine male rivalries and chauvinism, as Blavatsky recruited Henry Olcott, or as Catherine the Great recruited Potemkin. If she marries such a man he is likely to be merely an emasculated puppet. In contrast, charismatic men invariably enjoy the comforts of wives, families, supporters and sometimes even mistresses, who do not feel resentful or emasculated when given an order.

The charismatic woman also faces narcissistic insult almost daily. If she answers the phone the caller may assume that she is the receptionist and ask to speak to the manager, assuming this to be a man. Rude males who do not know her, feel free to speak over her or to interrupt. If she behaves aggressively she may be disliked rather than admired (Bass & Riggio, 2006). If she has children they may judge her harshly and blame her for missing out on their mothering because of her busy schedule, whereas if a father is unavailable because he is making his way in the world, then that may be more acceptable. Until she has made a name for herself she is not accorded the same respect that men receive when speaking publicly, and even when she is, her words are best received when restricted to domestic and related topics. Her pronouncements on politics, conflict and war are sometimes not taken very seriously because she is not perceived as one who may have to walk the talk or take a risk, but as one enjoying noncombatant status.

However, again there are exceptions, and perhaps none more so than Elizabeth I who, in her speech to her troops at Tilbury Fort in 1588, both acknowledged and embraced her status as a woman, while turning this into inspiring rhetoric — words a soldier might die for — when she proclaimed:

> Let tyrants fear: I have always so behaved myself that under God I have placed my chief strength and safeguard in the loyal hearts and goodwill of my subjects. And therefore I am come among you at this time not for my recreation or sport, but being resolved in the midst and heat of battle, to live and die among you all and to lay down for my God and for my kingdom and for my people, my honour and my blood, even in the dust. I know I have the body of a weak and feeble woman, but I have the heart of a King, and the heart of a King of England too! (Weir, 1999, p. 393)

The charismatic woman may also be threatened sexually in ways that charismatic men are not. The powerful man remains sexually desirable as he gets older, but the charismatic woman is always vulnerable to the younger, prettier female no matter how powerful she becomes. The charismatic woman usually does not have a retinue of males eager to have sex with her (Catherine the Great being an exception here; Troyat, 1994); however, this may not be a bad thing — the sexual availability of women to the charismatic male may enable him to further indulge his narcissism, as did Mao Tse-tung (Li, 1996).

None of this is much different from the experiences of ordinary women every day, but the narcissistic personality needs and demands much more than this, and daily she does not receive it. Hence charismatic women are often not in good psychological shape — Melanie Klein, for example, who commentators and biographers routinely describe as a 'powerful though tragic figure' (Stevens, 1998, p. 88; Grosskurth, 1995).

However, the question asked — Why are there so few charismatic women? — may not address the full range of possibilities associated with charisma. Men express their dominance through overt power and/or achievement, but women tend to do it more indirectly through relationships, and this may lead to quite different results. Further, narcissism has been defined in almost totally masculine terms. Having a sense of self-importance, fantasies of success and power, limited empathy, an inflated sense of entitlement and even a degree of arrogance are all widely considered to be part of the traditional male role, at least in much of the world. It can also be shown that several other personality disorders are gender-biased, and therefore what is labelled narcissistic personality disorder in men, may manifest differently in women, perhaps as histrionic personality disorder. But both of these may be merely gender-stereotyped expressions of the same underlying pathology. This argument has been made by Theodore Millon in his text, *Personality Disorders In Modern Life*,[1] and some associations have begun more recently to be found between histrionic traits and charisma (Khoo & Burch, 2008). Too little is known to settle this matter, but the answer to the question of why there are so few charismatic women may turn out to be that there really are many who escape our gender-stereotyped views because they express themselves differently to men. Perhaps a good recent example might be Wendi Deng Murdoch.

The current wife of media mogul Rupert Murdoch, Deng Wen Ge was born in rural China, the daughter of a minor Communist Party official. Expressing her ambitions early, she Anglicised her name to Wendi in her teens. She began studying medicine in China but dropped this when an opportunity to move to the West arose. Through study, socialising and marrying up she managed to migrate to America where, seven months after receiving her Green Card, she divorced her first husband and sponsor (having previously lured him from his first wife and two children). Eventually she went to work for the Murdoch conglomerate, and in time she married the boss despite a 37-year age difference.

Deng has immense charm, and the happy knack of disarming others and wooing them to her side. She also has 'presence'; she has been described as melting the defences of hardened senior executives because 'she has no fear' (Ellis, 2007, p. 36).

> There's a certain amount of guileless guile about her ... she'll set her mind on something and the way she'll go about it is with a sledgehammer. She's not a genius, she's a sweetheart, she's a party girl, she loves it when everyone is having fun — she likes to facilitate that. That's what she does. (Ellis, 2007, p. 37)

She has adopted several personas over the years; her volleyball coach back in China recalled her as being 'a calm girl, not very talkative', but a current friend says that she:

> ... talks a mile a minute, and I know sometimes she would like to slow down, but I think there is so much she'd like to get done in a day ... that everything comes tumbling out. I adore her because she is very thoughtful and quirky and brilliant in a completely original and unpretentious way ... She doesn't put on false airs. (Ellis, 2007, p. 38)

It is this combination of the free spirit with what the Chinese describe as 'a calculating heart', an 'incredible ambition coupled with (a) lack of self-consciousness' (Ellis, 2007, p. 34), that suggests something akin to the charismatic mindset. There is also an occasional telltale social inappropriateness to her, indicating one whose thinking is slightly out of synch with those around her. An associate observed:

> I've been at a dinner party where she'll just blindly jump into the conversation and say something harmless but completely unrelated to the table discussion. Rupert (Murdoch) will indulge her, laugh it off, almost in the

> same way that you'll indulge a child, and Wendi doesn't
> pick up on that. (Ellis, 2007, p. 38)

In true charismatic style Deng seems to have been economical with the details of her background, including the circumstances of her first marriage. She is highly intelligent, with an MBA from Yale; she is confident, persuasive, fearless, an excellent communicator and an adroit social manipulator, possessing a massive sense of entitlement and high energy.

Hence, unlike the charismatic man, the charismatic woman may actually have two options to express her ambitions. She may choose to openly compete with men in the pursuit of power, or she may work her ways more subtly through relations with men, recruiting a successful man and advancing herself through him. If this is the path she chooses, she is likely to not be considered by others to be charismatic. But if the example of Wendi Deng is considered, then perhaps the case can be made that such individuals are charismatic in some crucial, and essentially feminine, ways.

The charismatic woman also enjoys certain advantages denied to the charismatic man. She can exploit the maternal role, turning on the charm and caring in a controlling, invasive way. A man in this study told me that the female charismatic leader he was associated with became almost gleeful after he had had a minor accident (he fell off his bicycle), and needed medical assistance. 'She sprang to her feet and took over with a gloating eye as much as to say, "Now I've got you"'. Charismatic women may also be more easily forgiven gossip and manipulation, and they are able to make greater use of small intimacies such as touch and infantilisms. Thus, despite the disadvantages for women inherent in charismatic leadership, some do arise. The narcissistic personality really has little choice but to strive to be special, regardless of the obstacles that lie in her path, and there are always tools available to a determined woman. But another reason why there appear to be so few charismatic women lies in women's use of particular psychological defence mechanisms.

Psychological theory has it that in normal development the young girl gains her sense of gender identity from identification with her mother, and the maternal goodness that the mother provides and symbolises — nurturance, compassion, empathy, sensitivity, service, self-sacrifice and so forth. Thus, a girl's identity is derived from seamlessly connecting to her first love object, her mother (in contrast to the boy, who develops his gender identity by separating at a tender age from his mother and identifying with his father). Because of the girl's continuing close connection with her mother, she comes to use introjection as a defence much more than males do, incorporating into herself her mother's sense of womanhood and making it her own. Introjection involves internalising an external attitude, trait or behaviour that is modelled by someone else; the receiver does this so thoroughly that it becomes part of their own self, something automatic and seemingly natural. In identifying herself with her mother, the developing girl introjects the 'good' values associated with the mother and thus secures her gender identity (MacWilliams, 1994, p. 239). By identifying herself with the good, the girl comes to be seen as valued both by society and by herself, in time becoming the virtuous woman so potently analysed by Kate Fillion (1997) in her book *Lip Service*. She does not compete with males for power because she inhabits a different realm that is above and beyond competition for power, although she competes with other females for status.

This probably places limits on the girl's inclination to rebel and to challenge authority, or at least redirects it towards more covert means. That many women do challenge authority (due to the combination of patriarchal social pressures and the opportunities of an egalitarian society, every modern woman eventually finds herself in situations where she must challenge authority), shows that this early adaptation is not definitive. It becomes overlaid with many other forms of socialisation. But it may be sufficient to inhibit the development of the kind of grandiose, all-or-nothing, risk-taking qualities taking that charismatic men exhibit.

Nancy Choderow, in her book *The Reproduction of Mothering* (1999), builds on this theory to explain why women in relationships tend to fear abandonment and men tend to fear engulfment. In Choderow's account, the man has already been through separation from his primary love object (his mother) and this was so painful that he never wants to risk experiencing anything like it ever again. So he prevents himself from getting too close to a woman because it risks the terrible pain of loss. A woman has no such fears. She separated from her primary love object at a much later stage, and presumably when her attachment to her secondary love object (her father) was more secure and supportive, as were her other family and social relations. Hence, in marriage, the woman moves ever more towards attachment and may even be quite happy to risk dependency because it holds few fears for her, whereas the man needs to detach himself from his wife and put some emotional distance between them in order to maintain his sense of himself. If Choderow's theory is correct it would also help to explain why so few women develop charisma, and why those who do develop it tend towards a masculine valance in their personalities.

Charismatic women also belie certain basic ideas about narcissism. The notion that narcissistic personalities cannot acknowledge vulnerability or insufficiency has to be modified in light of the emotional displays resorted to, at times, by charismatic women, for whom tears and the acknowledgment of need are more acceptable than for men. That she sometimes exploits her emotions for effect merely shows the consistent pattern of canny impression management that characterises all charismatic personalities.

The charismatic woman can usually be distinguished from strong noncharismatic female leaders such as Margaret Thatcher and Helen Clark, by the personality cult that she weaves around her, most typically Evita Peron ('I will return, and I will be millions'). However, the female cult tends to be self-limiting because it remains oriented towards feminine concerns of status, beauty, family, health and relationships; by far the greatest numbers of charismatic women are located in service and religious organisations, whereas charismatic men tend to be concentrated in corporate, political and military organisations. This is probably because in her regressive fantasies the female charismatic adopts the dual roles of mother and queen. Like teenage schoolgirls in the playground, her moment-to-moment desire is to be seen as the most popular and beautiful woman, an indirect source of power, and largely aimed at other women. In contrast, the regressive fantasies of the male charismatic are to conquer the world, to be both warrior and saviour, and his moment-to-moment desire is to prove that 'Mine's bigger than yours'. Perhaps feminine narcissism really is less about power, and more to do with social desirability and being perceived as a desirable love object. Perhaps power means different things to men and women, or at least to many of us.[2]

In sum, there are many powerful forces, both external and internal that impede, although they may not prohibit, the rise of charismatic women. It is hard to avoid the suspicion that, culturally at least, power is a masculine game. Despite her narcissism and genuine talent, the charismatic woman's mentality does not sit easily with cultural stereotypes, or the practical realities of power and authority; the world just is not structured that way. Yet paradoxically, at least in the corporate world, women leaders may actually be more transformational than male leaders (Eagly, Johannesen-Schmidt, & van Eigen, 2003). Clearly, the topic of female charisma deserves much closer study, and a useful place to start is with a case study of one such woman.

Germaine Greer

Every social transition needs its polemicists and role models, and the most outspoken role model of feminism has been Germaine Greer. The public image of Greer is of the former writer for avant-garde satire magazine *Oz*, author of the immensely successful *The Female Eunuch*, subsequent firebrand activist for women's issues, and latterly author of several controversial books espousing radical positions on society, art and gender. Brilliantly polemical, Greer is seldom far from the headlines, the lovable yet temperamental prima donna feminist fighting the good fight with a heart of gold.

It is her humour that most attracts. Some day someone has to produce a book titled 'The Wit and Wisdom of Germaine Greer', for hilarious stories of her abound. There is usually a feeling of liberation to these yarns (apocryphal though they may be), of release from stale constraint. For example, on overhearing a woman academic lament, 'I lost my mother last year', Greer airily opined, 'Mmm, sounds a mite careless to me'. Her lack of sympathy is not the point so much as her spontaneous irreverence. Everyone who has ever felt shackled by convention yearns to have the nerve, the wit and the creativity of Germaine. There is an almost magical quality to her, something quite vivacious and lovable, an immense elevation combined with an earthiness and humour that transports her far beyond the limited visions of the crowd. This is her function; she serves as a beacon of freedom for others.

However, this public image was qualified in 1997 when the biography *Greer: Untamed Shrew*, by Christine Wallace, appeared. Wallace did not so much challenge the image as reveal the reality that underlay it, and the self-promotion driving it. The facts turn out to be very different from the image. It seems that Greer was never really much of a feminist and had little early interest in women's issues. A close reading of *The Female Eunuch* and her other books reveals many antiwomen attitudes, including blaming women for male violence.[3] She is dismissive of the suffragettes and of women artists, has savagely criticised other feminists (Betty Friedan is 'crazy', Gloria Steinem is naïve and 'food-phobic'), and she opposed the Equal Rights Amendment in America. She has at times been quite contemptuous of other women, as in the 1970s when she called herself a 'starfucker' saying that she only had sex with rock stars who were her peers, unlike the 'slag heap' of females who come from 'the audience side of the footlights' (Wallace, 1997, p. 297).

If not at heart a feminist, Greer has nevertheless used feminism as a vehicle to live a spectacular public life. Her books are not about women so much as they are about herself and the life stages she has passed through; youth and sexuality in *The Female Eunuch*, motherhood (or her lack of) and the family in *Sex and Destiny*, menopause in *The Change*, her relationship with her father in *Daddy, We Hardly Knew You*, her relationship with her mother scattered throughout much of her writing, and her somewhat antiwomen interests in *The Obstacle Race* and *Slip-Shod Sibyls*. A critical examination of her output gives the impression that she does not like her own gender much.

The woman who emerges from Wallace's account is not the lovable, entertaining Germaine of the media. She is witty but not really warm, and her humour can have a harsh edge. She is successful, but also tough and driven. She is elitist and socially inappropriate in a way that suggests a lack of empathy with lesser mortals. She can be unreflective, self-absorbed and flexible with the truth. Behind her public image Greer appears most of all to be powerfully ambitious.

Yet she is so witty in public performance, and so iconoclastic in her writing, and so adept is she at presenting herself as a progressive spirit, and so disarming is she, that she remains widely loved. Thus the central question regarding her becomes, how does she get away with it? In a series of inconsistent and confusing publications Greer has assaulted every sacred cow of feminism, (Beatrice Faust calls her a 'Quisling'; Wallace, 1997, p. 329), yet she remains an icon of the movement. She has abandoned or contradicted most of her own radical pronouncements, yet everywhere she continues to draw an audience. Despite poor reviews, her books continue to sell in large numbers to a popular audience. She is surrounded by enormous goodwill despite having alienated many who have admired her. So protective is this goodwill that even biographer Wallace, in an other-wise unsentimental study, still could not quite bring herself to raise the question of Greer's mental health. Yet appreciating her psychol-ogy has to be the key to understanding the force of nature that is Germaine Greer.

As Wallace tells it, Greer's psychic development began with a loathing for her mother, whom she perceived as having turned her father against her. Her father went off to war when Germaine was only three, and he returned traumatised three years later. In this time her mother may have had an affair, and when the marriage resumed there were inevitable difficulties. Despite the problems, the marriage survived and two more children were born. But it seems that Germaine's demanding personality and the three-year separation from her father kept her permanently peripheralised within the family. Throughout her writings she sprinkles many criticisms of her mother, and in *Daddy We Hardly Knew You* she reveals that her father never even mentioned her in his will. It seems that Greer's life drama involves an enduring hostility for the female, and an unrequited yearning (but with a degree of resentment) for the male.

As described above, the developing girl gains her sense of gender identity from identification with her mother and the nurturing goodness that the mother represents. Hence her gender identity is derived from connection, and it is developed by introjecting the maternal values that the mother expresses and symbolises. But such connection, identification and introjection can lead to problems in adolescence and young adulthood, when the young woman needs to separate herself emotionally from her mother and to rebel in order to become her own person (it may be no coincidence that this is the prime age for interest in feminism). A woman who is closely identified with her mother may experience intense guilt when rebelling against her, (and at least some in the women's liberation movement were very concerned with this rebellion).[4]

It seems that because of her antipathy towards her mother, Greer was never inducted into this near-universal feminine pattern. She never grew up feeling intimately connected to her mother, never introjected her mother's values and world-view, never identified deeply with her. Or rather, if she did it was in a counter-identified manner; that is, she cultivated in herself those traits and behaviours that were the opposite of the ones she saw in her mother.

Where the mother–daughter relationship is problematic, there is an increased likelihood of the daughter becoming 'masculinised,' and rejecting the virtue that the mother symbolises; that is, assertive and oppositional rather than nurturant. Because of her lack of connectedness, Greer has exemplified to women struggling with separation and identity how they too might rebel against maternal and all other kinds of authority. This also explains why she has never strongly rebelled against men, for men have seldom been an oppressive force in her life (her book *The Female Eunuch* was suggested to her by a male editor). Wallace got it precisely right when she concluded: 'This is the key to why she has been an inspiration to so many other women. She has never surrendered her sovereignty. Germaine Greer was never tamed.' Or, in the jargon used herein, Germaine never surrendered her primary narcissism; she remained a free and wild spirit. In sum, it is her charisma that she is loved for, her radical autonomy, her ability to just be herself, and to doggedly not give a damn.

So how does she get away with it? In part because her appeal remains mostly restricted to women and they easily forgive her inconsistencies and her antiwomen statements. Such bitchiness is a daily part of women's lives, and they are simply applauding a winner. They are also inspired by her freedom, and they probably sense her vulnerabilities. Perhaps they also intuit that they (her audience) are her first love, maybe her only love, and that she needs them as much as they need her, and for the same reason — to fulfil some missing part of herself.

Also, to her credit, Greer has remained aloof from the extremes of feminism. Probably the single biggest factor in the decline of feminism has been the alienation of reasonable people by radicals espousing models of gender relations that bear no relation to daily life. For, despite the claims of Susan Brownmiller, Andrea Dworkin and others, society is not awash with rape, and marriage is not a form of prostitution, nor are separatism and lesbianism desired options for the vast majority of women. Further, where feminist theory has been applied to social problems it has failed miserably, typically in the issue of domestic violence (Dutton, 2006, 2007). Greer has wisely kept her feet on the ground and pitched her books towards the mainstream, and thus she retains her following.

But if Greer were asked how she gets away with it she might reply, 'Get away with what?' For, like most charismatic personalities, she seems genuinely ignorant of her inconsistencies and contradictions. In January 2005, she participated in the British television show *Celebrity Big Brother*, despite having previously said that such shows were shameful; her appearance was for a £50,000 fee. After only a few days she left the show to a storm of media criticism (some of it from her feminist sisters) that she was merely seeking publicity (Lumley, 2005). This did not bother her. In an explanation published in *The Sunday Times*, she simply said that she left the show because of its bullying rules (the other participants presumably were comfortable with these).[5] She seemed not to comprehend how the public might see her as a publicity-seeking opportunist motivated mainly by the cash. For all her worldliness she appeared to be quite naively unaware of the reasons for their censure.

This is probably because the narcissistic mindset is so self-absorbed that it misses the nuances of intimate relationships. Greer probably does not know how others see her, or, if she does, she feels no emotional response because she is not psychologically connected to them. She is massively impervious to criticism, as evidenced by her indifference to the mostly negative reviews of her recent books. If your ego is that big then you do not reflect on how others see you, so long as the applause keeps coming, and you do not suffer bad feelings when criticised. In her lofty disregard of public opinion, Greer may actually be revealing her true feelings towards a world with which she has been unable to involve herself intimately. In sum, she gets away with it by sheer force of ego and will, and a vast self-belief.

Because she has spoken so passionately for women, there is a perception of Greer as having herself overcome some adversity. Not so. Although her family had its emotional problems and was never wealthy, it was basically a happy home, her parents were highly literate (her father had an editorial job on a newspaper), she attended private schools and she received a generous living allowance from the government to attend university. Yet, despite her background of comfortable security and culture, there is a sense of righteous anger to her, analogous to a working-class chip on the shoulder, a 'surplus trauma' that is out of proportion to whatever sufferings she may have had. This is typical of many charismatic personalities; they seem to genuinely feel personally aggrieved, as if they have been cheated of their birthright. Frequently self-pity, intense ambition and a powerful sense of entitlement accompany this, as if one is an 'exception' as Freud put it. There may also be 'divine discontent', a ceaseless, creative yet oppositional searching for belonging, healing and home, expressed as a utopian project and an underlying rage expressed as attraction to conflict.

Surplus trauma derives more from the narcissistic mindset than from any objective conditions; it is rooted in psychological rather than external history. This is the origin of narcissistic rage, a rage not against objective disadvantage such as poverty or sexism, but at some loss that is felt to be much deeper and more personal, as if part of one's very self has been usurped. Greer's anger probably arose within her family during her formative years, and became structured into her personality as one of the building blocks of her ego. Because it is so primal it is beyond the reach of reason and argument (and of therapy), nor can it be healed with narcissistic supply, although this may ease it somewhat. In contrast, objective disadvantage can be healed; the poor boy makes good and the woman facing sexism overcomes through education, hard work and creativity. But surplus trauma is deeper and permanent. This is the psychology of the activist; the anger expressed on behalf of others is not the real anger.

To her great credit Greer has done the best one can with such a legacy. Her fulminations against injustice enable her to vent her rage in a prosocial way, while retaining a positive self-image. Although her rage has no relation to objective conditions, it is expressed through these, mostly with positive effects in her case, but the original injustice that was done to her remains unhealed. Her repetitive seeking-out of conflict is a re-enactment of the original conflict that drives her; it is her effort to heal herself, and to recover some lost part of herself: if she can right this wrong she can right the original wrong. Occasionally her rage leaks out undisguised, as in her piece on the death of naturalist Steve Irwin (Langton, 2008). This was no lapse; it was something as basic to her make-up as any of her provocative witticisms. It seems that a sense of personal grievance is necessary for charismatic development.

Daddy We Hardly Knew You is Greer's best book and her most self-consciously 'literary' work; ('The few thousand houses nestle in the river valley like sugar crystals in the cupped palm of a hand'; Greer, 1989, p. 43). Yet despite its subject, it is descriptive rather than psychological, and large chunks have little to do with her quest to identify her father's origins and motivations. At its best her writing is quite wonderful, and her description of the siege of Malta during World War II is superb. She has the happy knack of finding the revealing detail that brings events to life; she might have been a tremendous action writer. Usually, she is not so effective when writing subjectively; (Wallace quotes a friend's description that, 'She's about as introspective as a sweet potato'; Harrison, 1992, p. 88), and

Daddy We Hardly Knew You is a subjective work. But like most things Greer sets her heart upon it is a great success, a moving triumph of her writing craft and her thwarted relationship with her father.

Yet this leads to a consideration of the tragic dimension of Greer's life, for despite the applause she seems not to be a happy or fulfilled person. She has been unable to achieve an enduring loving relationship, and she is childless. In an interview in 1998 when she recanted her earlier stance on sexual liberation she admitted, 'Maybe I've never been sexually awakened at all' (Fallowell, 1988). Her needs for attention and control, for fame and fortune and to provoke and conquer, and her ability to satisfy these, seem to have blocked any deeper need for connection. Her strength and intelligence may have insulated her against reality, enabling her to avoid facing painful personal issues, yet failing to lift her out of herself. Success has come easily for her, perhaps too easily, and like most charismatic personalities she seems reluctant to confront herself.

Beyond her obvious talents and brilliant intellect there seems to be not a lot to Greer. The life she is living now, and the things she is doing today, look very much like the things she was doing 30 years ago. There appears to be a lack of growth and development in her life. While possessing great common sense she nevertheless seems to lack a template for the 'normal', especially in her personal life. She is realistic and acutely reasonable in her writing (despite her lateral forays), and well-balanced amid feminist fads, yet she seems incapable of those routine compromises that make the mundane viable, and that enliven community and relationship. She must be the headline, and cannot abide the footnote. She is probably not bitter, as some critics have claimed, she is too grandiose for that, but she may have become cynical. She has been content to recycle the same kinds of arguments about gender and society, altering her message here and there but keeping the basic story-line going, through a range of fairly predictable though always iconoclastic publications. Her photo-essay *The Beautiful Boy* could be interpreted as a counter-punch to those quasi-pedophilic David Hamilton photos of warmly-focused prepubescent female nudes (Greer, 2003). One wants to cheer her on while also recognising the limitations of her works. She continues to 'get away with it' just by being who she is. Like all charismatic personalities, her appeal is symbolic and rhetorical, and is undiminished by contradictions or inconsistencies.

Greer's writings are polemical rather than scholarly, and while she appears to hope that her recent book *Shakespeare's Wife* might finally bring her scholarly recognition, early reviews have not been promising (Hunt, 2007). She is most effective at the next level down, the public intellectual; in this arena she has earned her applause. At the height of feminism Greer stood out for her readiness to fight publicly with men, her plain enjoyment of doing so, and her frequent besting of them. This was singular; most women of that time were uncomfortable doing this. Nowadays she ventures far beyond her areas of expertise, sometimes concocting arrant nonsense; an article she wrote for *BBC Magazine* in 1994 drew an embarrassing torrent of rebuttal from music experts, and her 2008 essay 'On Rage' saw her accused of racism by some Aboriginal leaders (Wilson, 2008). She is not racist, but her vast self-importance inclines her to overreach. This was nowhere clearer than in her remarks about Michelle Obama's inauguration dress.[6]

The gap between Greer's actual accomplishments and her continuing popularity among feminists despite her antiwomen remarks, reveals something of the essence of charisma — its mesmerising allure. For Greer's appeal is irrational; she is not loved for what she says or does, but for what she is! She is loved for her being, which many women (and men) find intrinsically liberating and inspiring. She could stop writing altogether and she would still be adored. Her supporters are excited by her freedom, her wilfulness and her elevation. In her they see what they themselves were like before reality got to them, and what they might be like now if they had not compromised with the social world, and had not accepted the inhibitions and restraints of society. The hypnotic appeal of charisma is not based on accomplishments or words; it comes from the recognition by the followers of their own lost narcissism, and of the price they have paid for their 'normality'. The charismatic personality resonates with the followers' unlived lives, evoking an intoxicating yearning for what once was and still might be, and the leader does this just through his or her being.

Feminism deserves to be taken seriously but critically, and one's evaluation of it will likely depend upon whether one views it as a cause or an effect of the great social changes that erupted in the 1960s. With the benefit of hindsight we can now see that by the 1970s, women's equality was an idea whose time had well and truly come. Contemporary life with its computers and e-mails, microwave ovens and cell phones, designer labels and cafe lattes, boutique wines and international cuisine, huge homes, tiny families and double incomes, as well as the transformation of major industries such as tourism, hospitality, fast food and day-care (mostly staffed by part-time female workers), all this could never have been based on the gender relations of the 1950s. The complete equality of women in society and their full participation in public life constitutes one of the largest social revolutions in history, with implications that are still not fully clear. But like all great historical processes, the changes occurred not because of a handful of activists like Greer, but because of wider historical, economic, social and cultural imperatives, and because of technological innovations such as the contraceptive pill, and because of broader male sympathies, especially those mostly male legislators who passed reforms to deliver full equality to women.

Hence, the women's movement cannot really be likened to other liberation movements. Many people died fighting apartheid in South Africa, and in the 1960s cities burned for the cause of black liberation in the United States, but no one died in the modern women's movement. It drew on the support of a progressive majority of both men and women, and before it arrived women already had full political rights; the term 'liberation' is a misnomer — it should be 'empowerment'. In contrast, it is sobering to consider a women's movement that really did cost lives, the 1927 'de-hijabisation' program by communist authorities in Soviet Turkestan. In this attempted social revolution, 87,000 Uzbek women publicly repudiated their 'black cowls', and in the process 300 were murdered by male heads of Muslim families for doing so (Warraq, 2003).

In sum, it was not the activism of feminists like Greer but the collective sympathies of progressive people of both sexes that delivered equality to women in contemporary society, and enabled their full participation in public life. The women's movement was a revolution of cultural rather than political significance. Despite often claiming credit for this great transition, what feminists really did was to vocalise it, to model and symbolise it for their followers, and to create a supportive myth that sustained them through these changes.

Greer is still doing this. In 2004, I attended a talk given by her in Melbourne.[7] She drew a large crowd that included academics and young people who would not have been born when *The Female Eunuch* was first published. There was also a sprinkling of men. Her subject was Shakespeare's women, and she spoke passionately of Juliet's assertiveness and Kate's shrewishness. She was well received. A large part of her audience however, was bookish older women who appeared to be singularly passionless. It seemed that once again Greer was providing a role model for admirers who found it difficult to fashion liberating experiences for themselves.

Endnotes

1   Millon et al., 2004, p. 83. Millon also cautioned that, 'There are problems on multiple levels. Our conceptualisations are fuzzy, our samples are biased, our measures are biased, and our clinicians fall prey to their own biases'.

2   Needless to say, the mentality of a Catherine the Great (Troyat, 1994) was utterly different.

3   The actual quote is: '[Women] act as spectators at fights and dig the scenes of bloody violence in films. Women are always precipitating scenes of violence in pubs and dance halls. Much goading of men is actually the female need for the thrill of violence' (Greer, 1991, pp. 354–355; Wallace, 1997, p. 198).

4   Wallace unwittingly gives a revealing example of the closeness of this identification, and the sense of oppression that can accompany it, when she writes of 'the tyranny of Mills and Boon' (Wallace, 1997, p. 197). It is an odd notion, utterly foreign to masculine thinking. That something as passively impotent as a Mills and Boon novel could take on tyrannical dimensions shows how thorough-going can be the introjection of the 'good'.

5   *The Sunday Times* (16/1/2005).

6   'Greer ridicules that dress'. *The Age*, 20–11–2008, p. 3.

7   Melbourne Town Hall, September 16, 2004.

CHAPTER TWELVE

The Charismatic Alliance:
Franklin and Eleanor Roosevelt

*People love others not for who they are but for how they make
them feel.*

Irwin Federman

Because of the solipsism inherent in narcissism, theory suggests
that alliances between charismatic personalities and others
ought to be rare, and due to exceptional circumstances. This is
so, and only a handful of examples spring to mind — Catherine the
Great and Potemkin, Franklin and Eleanor Roosevelt, Evita and Juan
Peron, and Bill and Hilary Clinton. In each of these cases the alliance
occurred between a charismatic and a strong personality. Could an
alliance between two charismatic personalities occur? This is the least
likely possibility, but even this may happen. It is most likely when
one works for another, say, the junior working his way up in a corporation led by his boss, and it does not usually last for long.

However, in *Prophetic Charisma* I unearthed one instance of two
charismatic personalities at the peak of their abilities working alongside each other over several years. The circumstances were
exceptional. They were leaders of the pacifist movement in New
Zealand during World War II, and they were so beset by the authorities, who regularly raided their communal farm and rounded up and
jailed their followers, that they had little choice but to work together.
They continued their alliance for a decade after the war, although
again their choices were restricted; their farm was probably the only

place both felt safe. Each of these leaders originally came from within the mainstream church; they understood each other well, and were probably examples of what Erich Fromm has called 'benign' narcissism (Fromm, 1964, p. 77). Thus they were presumably more able to subordinate their own grandiosity to an external ethic, (chapter 13 discusses benign narcissism further).

The example of Franklin and Eleanor Roosevelt is the charismatic alliance about which we have the best information. This case is doubly significant because Franklin was an example of Kohut's second variant of charisma involving 'fixated' grandiosity, rather than 'compensatory' grandiosity, which seems more frequent. There was no neglect or abuse in Franklin's childhood, quite the opposite, and he too was raised within the strictures of an overarching external ethic, which he struggled with all his life. If any charismatic personality was well-placed to subordinate his narcissistic grandiosity to a relationship with another it was he, and the effort was partly successful, albeit at a price, and only at a distance.

Both Franklin's parents were strong personalities. His father was an odd mixture of conservative and adventurer, an entrepreneur born into a wealthy family who also claimed to have served in Garibaldi's revolution (Ward, 1985, p. 31). He had a quality of dignity that led most who knew him to refer to him as 'Mr James', and he exuded an air of benevolent but unquestionable authority. He doted on Franklin, and although aged 54 when his son was born, he was very much a constant presence in the boy's life and a positive influence on him. 'Franklin never knew what it meant to have the kind of respect for his father that is composed of equal parts awe and fear', (Ward, 1985, p. 120) his mother later wrote. It seems that father and son were constant companions, the one empathically supporting and nurturing the other through the vicissitudes of childhood.

Franklin's mother Sara was strikingly beautiful, refined and regal, and also from a wealthy background (her father had traded opium into China). Although 26 years younger than her husband, they were clearly very much in love, and she bore her only child — Franklin — when she was aged 28. Unfortunately, the birth was complicated and nearly killed both mother and babe. She also experienced several major losses around the time of the birth; a much-loved uncle, her nephew Warren whom she had personally cared for, and her brother Philipe.

The result of this traumatic birth circumstance led to Sara and Mr James agreeing to have no more children, and perhaps even abstaining from sex for the rest of their marriage. Instead, they invested their love and disappointment in young Franklin, for whom 'everything about his infancy seemed calculated to make him feel loved' (Ward, 1985, p. 112). His parents held him constantly and accompanied him everywhere, and he received the best that wealth could offer. It has been estimated that there is a minimum 7:1 ratio of the number of retainers needed to raise a child of the aristocracy (Tomsen & Donaldson, 2003), and Franklin certainly had this, with cooks, cleaners, nannies, tutors and other workers abundantly at hand to serve his needs.

He was not casually indulged or spoiled, however; he was raised with strictures governing all aspects of his behaviour. Sara once commented that a parent's mission was to keep a child's mind 'on nice things, on a high level', so every moment of Franklin's day was supervised to this end, including his play. The mood of such supervision can be glimpsed in a remark Sara made many years later when asked if she had always thought that her son would grow up to be President. 'Never, oh never!' she exclaimed. Her ambition had been far loftier. 'The highest ideal I could hold up before our boy — to grow to be like his father, straight and honourable, just and kind, an upstanding American' (Ward, 1985, p. 124).

Hence, Franklin had a remarkable upbringing. He was an only child but he was never lonely because he never knew privacy. So suffocating was his bondage to both parents, especially to his mother, that one friend remarked that Sara 'would not let her son call his soul his own' (Chafe, 2005, p. 7). Both of his parental models were loving but emotionally self-contained. So complete did his own emotional self-control become that by age four, when his parents travelled abroad for three months, he never exhibited a trace of distress, although when they returned he clung to Sara 'like a little soldier' (Ward, 1985, p. 115). He also became a skilled liar; perhaps his only way of having any privacy.

One effect of such a childhood can be to cause one to be strongly defended against the experience of anxiety; to become 'unflappable'. Although he was somewhat shy and anxious when very young (perhaps an internalisation of his mother's anxiety resulting from the circumstances of his birth and her recognition that he would be her only child), in time he developed an air of polite gravity and learned to suffer adversity in silence, yet always with a positive emotional valence derived from his parents' devotion. As an adult, when faced with difficulties that would scare a 'normal' person, he always remained calm and, if he lacked a solution to the problem at the time, he always felt confident that one could be found. As he would tell America during the Great Depression, 'We have nothing to fear but fear itself'.

Two anecdotes from his childhood demonstrate the atmosphere of all-enveloping security that Franklin was raised within. The first occurred when he was three years old. The family was on their way home from England on the steamship *Germania*, when they encountered heavy seas. Suddenly the ship plunged beneath the waters; everything went dark and passengers started screaming. After a time Mr James soberly observed, 'We seem to be going down'. 'It does look like it,' Sara responded.

As water began pouring into the cabin, Mr James went to seek the captain. He could not find him, but he ascertained that the ship was foundering. A huge wave had broken over it, washing one sailor overboard, ripping lifeboats from their davits, and smashing passengers and crew thither and yon. The pumps were working as hard as they could be driven and the stewards were bailing by hand, but despite their efforts the water in the cabin deepened.

'I never get frightened,' Sara recalled years later, 'and I was not then.' As the water edged up to the hem of her fur coat she removed it to wrap it around her son. 'Poor little boy. If he must go down he's going down warm,' she told Mr James.

As she did this, Franklin noticed a favourite toy floating in the deepening water. 'Mama, Mama,' he exclaimed, 'Please save my Jumping Jack.' Sara reached over and saved that too. The ship did not sink; the captain had been momentarily knocked unconscious by the wave, but he soon recovered and regained control of the battered ship and managed to nurse it back to Liverpool. But Franklin's parents' imperturbability, his mother's warm coat and her rescue of his toy, were ideals of security in crisis that remained with him all his life.

The second anecdote occurred when Franklin was 8 years old, and is related by his mother.

> One day ... we noticed that [Franklin] seemed much depressed and bound, do what we would to amuse him, not to be distracted from his melancholy. Finally, a little alarmed, I asked him whether he was unhappy. He did not answer at once, and then said very seriously, 'Yes, I am unhappy'.
>
> When I asked him why, he was again silent for a moment or two. Then with a curious little gesture that combined entreaty with a suggestion of impatience, he clasped his hands in front of him and exclaimed, 'Oh, for freedom'.
>
> It seems funny now but at that time I was honestly shocked. For all he was such a child, his voice had a desperate note that made me realise how seriously he meant it.
>
> That night I talked it over with his father ... We agreed that unconsciously we had probably regulated the child's life too closely. Evidently he was quite satisfied with what he did with his time but what worried him was the necessity of conforming to given hours.
>
> So the very next day I told him that he might do whatever he pleased that day. He need obey no former rules nor report at given intervals, and he was allowed to roam at will. We paid no attention to him, and, I must say, he proved his desire for freedom by completely ignoring us. That evening, however, a very dirty, tired youngster came dragging in. He was hungry and ready for bed, but we did not ask him where he had been or what he had been doing. We could only deduce that his adventures had been lacking in glamour, for the next day, quite of his own accord, he went contentedly back to his routine. (Ward, 1985, p. 128)

Perhaps the key passages in this extract are 'we paid no attention to him', and 'we did not ask him where he had been or what he had been doing'. These remarks suggest an underlying parental message that the pursuit of 'freedom' may accompany a withdrawal of love. If this was the message that Franklin received, he had the skills to consciously avoid or deny the anxiety it produced, although it may have bubbled away beneath the surface. He may have felt a deep rage.

Raising a child in an atmosphere of all-enveloping love and security may lead to problems. Inexperience or denial of anxiety may lead to a lack of empathy for others who have to manage anxiety. Having all one's needs taken care of by others may reduce one's own problem-solving ability. Being constantly related to as if one is special may make one unable to relate to others as an equal. Such love and security is not necessarily a good thing, but for Franklin other issues soon complicated things.

Franklin was 8 when his father, aged 62, suffered a heart attack from which he never fully recovered. Now Mr James became someone to be cared for, rather than a carer. Franklin's dutiful response became to play the role of the bright and happy child, never difficult or fractious, for his father. He and Sara moved even closer together emotionally in their loving conspiracy to keep his father joyously alive. His adult buoyancy and incessant optimism probably date from this period, as may several other striking self-management strategies. Franklin learned when things went wrong, to merely withdraw for a while and then carry on as if nothing had happened. Given that things inevitably did go wrong — his father's subsequent heart attacks for example, about which he could do nothing — he learned to just keep on keeping on. It worked as well as any strategy might in the situation.

His father's illness had profound ramifications, such as the family's five trips to the German spa at Nauheim where Mr James went hoping for a cure. Here Franklin was surrounded dawn to dusk by sickness and impending death. He knew never to worry his father, nor to appear worried or repelled himself. He never spoke about these stays, but he must have managed by deliberately not noticing the numerous personal tragedies unfolding daily around him, while putting the most positive spin he could muster on the sights and sounds he encountered.

The final pillar of Franklin's childhood was the strong religious faith he inherited from his parents. Theirs was a muscular Episcopalianism that emphasised civic responsibility and eschewed display. Mr James was something of a reformer, strenuously arguing the case of the poor in his parish, while Sara imparted the notion of 'noblesse oblige', and both set examples of abstemiousness and hard work in all areas of life. Their faith was simple and honest, and Franklin always loved to sing hymns and to find solace within the walls of the rugged old church in which his father and relatives had been baptised. His faith may have been limited, but it allowed enough room for him to come to consider himself as being in some way 'chosen' (Davis, 1984).

There is a revealing story of the nature of Franklin's faith that was told many years later by his wife. Eleanor asked him one day if perhaps she ought not to impose their religious attitudes upon their children, but perhaps instead to allow them to make up their own minds as they grew. Franklin was reportedly astonished at the suggestion, and insisted that their children be trained as they themselves had been, as Episcopalians. 'But are you sure you believe in everything you learned?' Eleanor asked him. 'I never really thought about it', he replied, adding, 'I think it is just as well not to think about things like that too much' (Ward, 1985, p. 157).

Franklin never went to school until he was 14; he was educated at home by his parents, governesses and tutors. He was enrolled at Groton, a boarding school for young gentlemen, at age 12, but apparently both parents wanted to hold on to him as long as they could, so his departure was put off until he reached 14, and he began in Form Three rather than in Form One. This was not so exceptional; Mr James had also not gone off to school until 14, but most of his peers began at 12 in the first form.

Groton was not like the horrendous English 'public' schools of that time. It was warm and supportive, offering a church-oriented education in a home-like atmosphere. At bedtime both the Rector and the schoolmaster's wife personally said goodnight to every pupil (Ward, 1985, p. 180). However, it was at Groton that Franklin suffered his first great narcissistic disappointment. At home he had been surrounded by supporters and loving family who elevated him and inflated his self-esteem; he had been liked by everyone. At Groton he worked 'almost desperately to replicate that world' (Ward, 1985, p. 180) but he failed. He was just another pupil making his way as best he could through the playground pecking order. He was never popular, seeming to be always a little out of sync with the other boys, probably because his previously sheltered upbringing and the inflation of his self-importance that had been encouraged there, gave him an air of unreality. The charm and cordiality that had so pleased his parents now seemed artificial, even oily, to his peers. He compensated by becoming oppositional and argumentative, but that only alienated them further. He earned respect for his intelligence, but little else. He knew more than the other boys about the world beyond Groton, but they knew more about being boys. He came to see himself as being in some way cheated out of the exalted status that was his due, and he always felt like an outsider.

Franklin remained at Groton for four years, and then he entered Harvard where he studied law. Two months later his father died. As was expected of him, he hid his grief, consoled his mother, and in the New Year departed again for Harvard. Next summer he accompanied his mother on a holiday to Europe, and upon returning learned that President McKinley had been assassinated, and that his uncle Theodore Roosevelt was now President of the United States. Theodore was to become an influential mentor for Franklin.

However, his time at Harvard went no better than at Groton. He was an indifferent student, his grades averaging a high C, and again, without being actively disliked, he was passed over by the inner circles. The prestigious student club Porcellian refused him admittance, though his cousins and several friends were accepted. He later described this as 'the greatest disappointment of my life', a second narcissistic wound.

Again, it seems that his demeanour was the problem. He gained a reputation for insincerity and for sinuous opportunism, for inappropriate and superficial charm, and he was perceived as lacking manliness. This latter accusation cut very deep; he had been raised by Sara in dresses until the age of six, in accordance with some obscure custom of the time. Later, his sexual development was quite slow and he was taunted as a 'good little mother's boy' (Ward, 1985, p. 251). This appellation was somewhat justified as Sara eventually moved to Boston to more closely supervise him, and although she insisted that she would not interfere in his college life, she could not help herself. Franklin felt obliged to provide an amiable torrent of letters, assuring her that he was happy and doing well, but was much too busy to meet her as often as she wished. The stratagem worked; he kept her at arm's length with a mixture of charm and deception, but he was struggling to deal with a continuing series of disappointments.

Franklin's career at Harvard was modest. He never finished his degree, the positions he applied for he seldom gained, girls were unimpressed with him, and his only real success at Harvard came when he became Editor of the *Crimson*, the university newspaper. He poured his energies into this job, neglecting his studies and genuine appreciation followed, though it was still not great. He was later remembered as having limited ability, but being quite a good manager; one colleague recalling that 'in his geniality there was a kind of frictionless command' (Ward, 1985, p. 239). This position was a great learning experience with relevance for the rest of Franklin's life. He had instinctively found his way to the heart of the communications network in a large organisation, and he was now able to project the kind of image of himself he desired. He also became interested in power; his presidential ambitions date from about this time (Ward, 1985, p. 253).

But the anxiety over his manliness never left him. He had been referred to by his cousins as 'Miss Nancy' and 'the feather-duster', and such a primal slur probably accounts — at least in part — for his youthful decision to marry. This defused doubts about his masculinity, and also recruited an ally in his struggle to differentiate himself from his mother. He was 21 when he proposed to his cousin Eleanor; she was 19.

Born into a troubled family, Eleanor's father was the younger brother of Theodore Roosevelt. He was frail and anxious, an alcoholic, and he may have suffered from bipolar disorder. Her mother, although beautiful, came from a cheerless, dysfunctional family, and she had her hands full coping with her wayward husband. Consequently, Eleanor was farmed out to unsympathetic relatives for much of her childhood. She felt anxious and abandoned from an early age. She was routinely humiliated by those around her, and mocked for her lack of beauty and her seriousness (her mother nick-named her 'Granny'). Her father entered a sanatorium when she was seven, her mother died when she was eight, her younger brother died when she was eight, and then her father died when she was ten, after which she was again placed in the care of bullying relatives. Her cousin Corinne later remarked that, 'It was the grimmest childhood I had ever known' (Chafe, 2005, p. 9).

Eleanor was never playful or carefree. She coped by developing her nurturing side; giving to others, making herself useful, helping those in need and generally putting herself last. This defused most opposition and gave her a role and even a little pleasure; in later years she was to say, 'The feeling that I was useful was perhaps the greatest joy I experienced' (Ward, 1989, p. 55). But in time her adaptation to adversity gave her a greater sensitivity, and made her more substantive than her peers. When finally, at age 15, she was sent off to a finishing school in England, she found her niche.

This school was run by a sophisticated French woman, Mademoiselle Marie Souvestre, who doted on Eleanor and intro-duced her to art and philosophy. Eleanor soon became Souvestre's 'supreme favourite' and the idol of her fellow students. It was a period of unqualified, brilliant success, unexpected yet fully earned. Her teacher later described her as 'the warmest heart I have ever encountered', while Eleanor recalled that, 'Whatever I have become since, had its seeds in those three years of contact with a liberal mind and strong personality' (Ward, 1985, p. 303).

Back home Eleanor began to be courted by cousin Franklin. In some ways it was an opportunistic marriage for him; Eleanor was independently wealthy, she had higher status than him and she gave him access to President Theodore. There were also no rivals for her (he had been rejected by a previous partner). For her too, there was much to be gained from the marriage; a genuine family (she got on well with Sara), and avoidance of the arena of life where she felt most inadequate — romance and the social whirl. They seemed genuinely in love, but neither had any real understanding of what that involved; they were simply solving whatever problems they encountered as they went along, albeit she had a more romantic attachment than he did.

At 25, Franklin gained admission to the New York Bar. He left Harvard and, using family contacts, got a job as a lawyer. He was ineffective, but the job led to the first stirrings of what became a social conscience. He found himself enjoying court work and sympathising with the disadvantaged. He was competing and winning in a field where breeding counted less than intelligence, and where arguing the case of the needy almost always gave him a moral advantage over his peers among the rich and powerful. He began to change his values and class sympathies so he could advance himself by championing the underdog.

However, the 1905 marriage of Franklin and Eleanor was troubled almost from the start. Eleanor became pregnant with the first of their six children not long after the marriage, and gave birth to a 10-pound daughter, damaging herself in the process. She was an anxious mother, and Sara controlled every aspect of the household. Another even larger baby was born soon after, and then a third 10-pounder, all within five years of marriage. This latter child died when seven months old. A month later Eleanor became pregnant again. Their fourth child was born in 1910, and by now Eleanor was barely managing to hold herself together. There would be two more babies arriving in the next few years. Exhausted by children, dominated (sweetly) by Sara, and with little support from her husband, it was probably only the skills she had learned while surviving her difficult childhood that enabled her to endure. But Eleanor was not the kind of person to crack.

Franklin seems to have lived in something of a narcissistic fantasy in the first years of his marriage. He neglected his wife emotionally, and spent much time away with his friends or travelling. He mimicked the behaviours and mannerisms of his uncle Theodore, who had gained a Nobel Peace Prize for his role in negotiating an end to the Russo-Japanese war. He was self-centred and out of touch with common realities such as earning a living and being a husband.

Eleanor tolerated this because she was so self-abnegating, but she withdrew into silences and bitterness, which Franklin not only did not understand, he could not see why he should. He began to think of entering politics, trading on his uncle's name and following in his footsteps. When Eleanor was in her fourth pregnancy he spent still more time away, planning his entry into politics, sailing and picnicking and racing horses. He was 28, handsome, rich and adventurous, and Eleanor was writing pleading letters begging him to return home. She was becoming more dour and he was becoming more free.

Franklin entered politics as a Democrat. As a Republican he would have been competing with other sons of blue-bloods, and he had not fared well doing that, but as a Democrat he would have the advantage of his name being linked to the President, and he would stand out among the rank and file as a patrician. It was a smart choice, but it also meant that to survive in such a milieu he would have to broaden himself. He struggled to do this, but perhaps his social inadequacies were more excusable to his new Democrat comrades, who may have believed they arose from class differences rather than personal limitations. Again he was not liked at first, and again he was not one of the boys; but this time he had an edge. The example of his feisty uncle saw him defining himself as a progressive reformer, and his experience editing the *Crimson* gave him the expertise to project a positive image of himself in the press. As a New York State Senator he opposed the machine politicians of Tammany Hall, and they fought him, but eventually an arrangement was reached that saw him become a valued member of the club.

Supporting Franklin in politics was the last thing Eleanor had imagined herself doing when she married, but she did her duty and even began to enjoy it. Her role as a political wife gave her a break from childcare (which was delegated to others), and also a degree of independence from Sara. Later she was to write, 'For the first time I was going to live my own life … to stand on my own feet … something within me craved to be an individual. What kind of individual was still in the lap of the gods' (Ward, 1989, p. 135).

They were both thinking their ways through to their mature philosophies, coming from utterly different perspectives, but staying the course and growing from the challenges that arose. Franklin made a poor initial impression as a senator; he was accused of being shifty and hypocritical, opportunistic and flexible with the truth, which he often was. But he learned fast, and he even developed some insights into himself, including into a deep hostility that lay within him. By his own later admission, he was 'an awfully mean cuss' when he first entered politics. But most importantly, his thinking was developing (no doubt under some influence from Eleanor), towards a progressive vision of 'interdependence — our mutual dependence one upon the other — of individuals, of businesses, of industries, of towns, of villages, of cities, of states, of nations' (Ward, 1989, p. 171).

Franklin was not an impressive performer in any of his early political posts. But he developed the skills he needed, including making alliances with those who could serve him well. He also developed a gift for the kind of story-telling that could reduce complex issues to metaphors able to be understood by the person in the street. His lawyerly skill of overwhelming others with detail was also useful. The arrival of their sixth child (Franklin had always wanted six) in 1916 enabled Eleanor to put childbearing behind her, and their sexual relationship probably ended then as well (Ward, 1989). She had never enjoyed sex; her ideal of marriage was more about companionship and duty than sex, and she was unable to frolic or play with Franklin. He, however, was also uncomfortable with intimacy and emotion, and utterly unable to express his feelings honestly. They drifted apart.

These were the years of World War I, and in 1916 Eleanor took up war work to salve the disappointments of her marriage, to relieve the tedium of motherhood, and to avoid the social life that she found so intimidating. She was feeling her way forward by using the strategy she had always relied upon, making herself useful to others, but her biggest test occurred two years later when she discovered that Franklin had been having an affair with her former secretary, Lucy Mercer. This confronted Eleanor with her greatest childhood terror: the anxiety that those she loved would mistreat, betray and abandon her. Her ultimate fear had arrived.

Sara played a major role in sorting out this crisis by promising to cut Franklin off without a penny unless he gave up Lucy. He agreed to do so and he and Lucy distanced themselves, although they maintained a correspondence. Lucy remained the great love of Franklin's life and although she later married another, much older man, after this husband died she returned to Franklin's circle to comfort him in his dying days (hers was the last face he saw).

This was the last of Eleanor's illusions and the wound remained raw all her life. However, she was not one to remain moping forever, and the betrayal may even have liberated her because it freed her to become her own woman in ways she may not have if she had continued to feel constrained by obligations to Franklin. After the disappointments receded she realised that she had faced and survived her worst fear, and there remained much more life to live. He, in turn, bent over backwards to placate her, even attending church for a time and taking a more central part in the raising of their children. Together they kept up an appearance of marital commitment, and although intimacy had ceased between them, they now resumed a working relationship that empowered them both.

Franklin's love for Lucy matured him; a cousin recalled years later that it 'seemed to release something in him'. Previously he had looked 'at human relationships coolly, calmly and without depth. He viewed his family dispassionately, and enjoyed them, but he had … a loveless quality, as if he were incapable of emotion' (Ward, 1989, pp. 414–415). In his first year of marriage he had suffered nightmares, sleepwalking and a prolonged and severe outbreak of hives, suggestive of deep conflicts working through him. It seems that his childhood fusion with Sara had left him with a deep ambivalence. He had overt feminine qualities in his character, such as a fondness for capes and a predilection for designing elegant costumes for himself. His close adviser Thomas Corcoran once described him as 'the most androgynous human being I have ever encountered' (Ward, 1989, p. 552). At first he and Eleanor had shared an inability to be intimate and to trust deeply; his way of handling these issues had been to simply avoid them. But with Lucy it seems he achieved the breakthrough to himself that he needed, and from then on he felt more able to mix and mingle in human affairs without feeling like an outsider. He remained an outsider, of course, but he had tasted love and its healing power; he had become a much more authentic human being.

Franklin continued in politics and Eleanor continued to develop herself, taking up with women friends and becoming more political in her own right. She later wrote, 'I realised that in my development I was drifting far afield from the old influences … I was thinking things out for myself and becoming an individual' (Ward, 1989, p. 565). Her new friends were activists and lesbians, and she even befriended Louis Howe, Franklin's closest political confidante and a man with whom she had little in common. They shared an intellectual and personal bond around Franklin's career and Democratic Party interests.

Then at age 39, Franklin caught polio. He survived, but now, like Eleanor, he too faced one of life's ultimate tests. At first he retreated into denial and compensatory optimism, imagining that he was 'special' and that the rules of polio did not apply to him. But as time wore on and he made no recovery in the use of his legs, he was forced to confront the limits of his narcissistic illusions, and to see himself as a vulnerable, needy person bound to a wheelchair. He withdrew from politics and set himself up in a healing retreat at Warm Springs in Georgia, where he pioneered the then-new discipline of physiotherapy, becoming a guru-like figure to other 'polios' who flocked there to take the healing waters. He made Herculean efforts to rehabilitate himself, inspiring others with his buoyant optimism and calling himself 'Old Doctor Roosevelt'. He did not get the cure he hoped for but the effort strengthened both his mind and body. He beefed up his arms and upper body. His face broadened, its features deepened. He learned patience and found a willpower within himself that he might not have found otherwise. After five years, with the aid of special callipers clamped to his legs, he managed to 'walk' a few steps on the arm of his daughter.

This was the limit of all he became capable of physically, but the effort to survive and rehabilitate himself had transformed him psychologically. Before polio he had been something of a shallow lightweight, but now there was depth and substance to him. He had felt the bitterness of loss and despair, sensed his frailty, confronted death and seen through the narcissistic illusion that he was somehow different to others. Although he remained fundamentally a loner, he developed empathy for the rest of suffering humanity. When he re-entered politics — this time pressed into service by popular demand within his party, the opposite of his experiences at Groton and Harvard — he was a very different man to the New York senator of a decade earlier.

Franklin Delano Roosevelt (FDR) became the greatest American president of the 20th century, serving four terms during the Great Depression and World War II. His achievements are unparalleled in modern times. Before FDR there was no Social Security, no regulation of the stock market, no federal guarantees of bank deposits or of the right of workers to bargain, no minimum wage or maximum hours, no price supports for farmers or federal funds for electricity development, and no federal commitment to equal opportunity or high employment (Ward, 1985, p. 6). He also oversaw the expansion of the United States military from 16th in the world to first. Through sheer confidence, trial and error, picking the brains of others and weekly, optimistic 'fireside chats' broadcast on the radio across America, he solved the Great Depression (Ward, 1985, p. 8) before J.M. Keynes's published his 1936 *General Theory of Employment, Interest and Money* (Jenkins, 2003, p. 78).[1] His dictum that 'great accumulations of wealth cannot be justified on the basis of personal and family security' (Jenkins, 2003, p. 89), remains a beacon for progressive politicians everywhere; (it also saw him labelled a 'class traitor' by his peers; Jenkins 2009, p. 85). He looked forward to the creation of the United Nations, to Negro equality, and to the dismantling of colonialism.

Through all of this, Eleanor, though rarely standing literally 'at his side', also toiled mightily for the needy and downtrodden. Work became an 'addiction' for her (Chafe, 2005, p. 29), as she travelled America searching out injustices; she was away crusading so often that a Washington newspaper once ran the headline, 'Eleanor Roosevelt spends night at White House'. She became an effective organiser, and took up the causes of women and Blacks, war widows and poor farmers. From sharing the prejudices of her class and time, she became pro-Jewish and pro-Negro. She would return from her journeys filled with fervour for progressive change, proposing initiatives such as antilynching legislation that went far beyond the political realities of the day. Franklin found many of her proposals, such as the Wagner-Rogers Bill aimed at rescuing Jewish children from the Holocaust, to be politically unworkable (Baker, 2008, pp. 115–125), but he listened to her with special consideration. The days when she returned from her journeys were times they devoted exclusively to each other, with long relaxing talks and easy communion, the one area of their marriage that continued to run deep. They formed a team unique in American political history; he managing the vast matters of economy and war, yet respectful of her strength and attentive to her convictions that the little people also matter.

Quite possibly neither was capable of an intimate, loving relationship, but they seemed to intuit each others' needs, as if they were two sides of the same coin. Their daughter, Anna, once explained this by saying that Franklin had had, 'too much security and too much love', while Eleanor had not had enough (Chafe, 2005, p. 55). But the kind of love that Franklin had had too much of may be almost as problematic as not having enough. They had many difficulties; according to one observer, Eleanor never got 'accustomed to his lack of real attachment to people' (Chafe, 2005, p. 52). Yet through her and her causes he may have found the only kind of attachment to others that he was able to have. He was, according to another political acquaintance, the 'most complicated human being I have ever known' (Ward, 1985, p. 9).

Is it possible to generalise from the relationship of the Roosevelts to the nature of charismatic alliances? Despite the improbability and infrequency of such pairings, some observations may be made.

First, it was an alliance in which each worked in very separate arenas, and relied on the other to perform tasks that each was incapable of. Because of his infirmity, Franklin simply could not get around, whereas Eleanor could. Nor did he have the time or empathy to genuinely listen to the downtrodden, whereas again, she did. Equally, she could never have gained the kind of political power he did. This demarcation of competencies was probably one of the main factors that made their partnership successful. They cooperated, but not on the same issues, and they 'would each have been less than half of the total if they had broken the marriage in court' (Bishop, 1975, p. 17).

Second, their working alliance grew from the great personal struggles they shared. They had been little more than children when they married, with no comprehension of the challenges ahead. He broke her heart, although in truth it had been broken years previously. She stuck by him through his infidelity and polio. They were able to replace their lost personal relationship with shared political ideals. Somehow, through all of that, each continued to grow, and to measure themselves and the other, arriving at a hard-won alliance of respect, trust, need, appreciation and acceptance. There was no hubris in their relationship.

This stands in contrast to the much more opportunistic relationship of the Perons (despite Evita's cloying public adulation of Juan Peron), and also to the much less character-building lives of the Clintons. Perhaps the best partnerships of any kind are forged through adversity and genuine respect.

Finally, although Franklin's and Eleanor's partnership was incomplete, it is hard not to conclude that each gained greatly from the other personally, and even spiritually. Each was a great teacher and helpmate for the other in their shared quest for fulfilment (within the parameters of their formative influences). Perhaps it is from surviving some great adversity, and then forming loving relationships that heal and teach one humility and humanity, that the greatest wisdom comes.

In sum, the charismatic alliance probably works best when there is enough shared understanding and trust, and even love, for the two partners to always feel close, yet when there is also enough distance between them in some way, for the charismatic personality not to engulf or diminish the other.

Endnote

1   There has recently arisen a revisionist argument about Roosevelt's handling of the Great Depression, which holds that rather than solving the depression he actually lengthened it by misguided populist policies. Where this argument is not simply false or mischievous (see Chait,

2009), its proponents nevertheless seem to overlook that a depression is not solely an economic problem. It has a human dimension also. The Roosevelt years were actually a period of cultural renaissance for America, a time when Americans who felt their values were under threat, reasserted and rediscovered themselves, and this was partly a result of his funding for a wide variety of social programs including, for example, spending on the arts. Perhaps more astute economic policies than those implemented by Roosevelt might have solved the depression sooner; (contemporary Republicans have a point when they argue against Barack Obama's stimulus packages — it might be quicker and cleaner to just let debt-burdened industries die). But in both these cases it is likely that the human cost of doing so would be unacceptable. Roosevelt 'solved' the great depression by economic policies that worked well enough, while also connecting with the population in such a way that they maintained their belief in the American nation and system. Obama is attempting the same; no doubt some historian will emerge some day to argue that he got it wrong in important ways, as no doubt he must to some degree.

·····

CHAPTER THIRTEEN

· · · · ·

Cosmic Narcissism

The self in the psychoanalytic sense is variable and by no means coextensive with the limits of personality as assessed by an observer of the social field. In certain psychological states the self may expand far beyond the borders of the individual, or it may shrink and become identical with a single one of his actions or aims.

Heinz Kohut

What can be said of the psychology of the virtuous charismatic leader? Do such 'saints' really exist? Can their strengths be reduced to their weaknesses, or is their apparent health merely some pathology in disguise? Are they somehow miraculously pathology-free? Is it even possible to see past our own illusions, superstitions and prejudices regarding such astonishing individuals? What is the genesis of great virtue?

These questions hinge upon free will. As Dostoyevsky argued in *Crime and Punishment*, the truly good person must be capable of evil, yet willing to choose the good; anyone incapable of evil, say, a person so brainwashed and guilt-ridden as to be unable to do evil, is merely inhibited and there is no particular virtue in that. Or, the 'saint' may be so predisposed, through some combination of temperament and circumstance, to embrace virtue in a fairly limited or even naive manner, as perhaps Francis of Assisi and John of the Cross presumably did (accepting their mythologies at face value). While acknowledging the great merit of such individuals we also recognise their limitations; we do not invite them to become leaders in the greater world where the lives of many might depend upon their choosing some lesser evil over an impracticable good.

The virtuous hero who is able to live and die for others strikes us as truly extraordinary, inspiring us to emulate them as the child models itself on a parent. Because we need to believe in such people, culture and religion provide us with myths of great saints and heroes to measure our own behaviours alongside. These stories are seldom genuine, power really does corrupt, but this makes the small number of truly virtuous individuals even more remarkable. Their failings remind us that they really are human like ourselves, not free from pathology, yet somehow able to choose the good.

One of the most remarkable was Cincinnatus. In 458 BC, Rome was beset by two tribes, the Aequians and the Sabines, who penetrated nearly to the walls of the city. Two armies were sent against these foes. The first, led by Nautius had some success, but the second led by Minucius bogged down, and was near to defeat when a desperate message was sent to Rome pleading for assistance. Nautius was called for, but despite his limited success he was not considered up to the task. The Senate sought a more effective military leader and unanimously voted for Lucius Quinctius Cincinnatus, and appointed him the new dictator.

Cincinnatus was sent for and was found hard at work on his farm, unaware of the dire peril Rome faced. He returned to Rome by river, thus avoiding the hostile forces surrounding the city and he accepted the post of dictator, but only for a limited period of six months. Then he began to raise an army. He appointed Lucius Tarquitius, a popular figure who was reputedly the best soldier in Rome, as his cavalry master, despite the fact that Tarquitius was too poor to keep a horse and had previously served only as an infantryman.

After his army was ready Cincinnatus marched out and engaged the enemy. In a series of brilliant manoeuvres he won the day. He was generous to the defeated soldiers, allowing them to depart with their lives. However, he humiliated the leaders by taking them into slavery. He distributed the booty of battle fairly among his own soldiers, but he neglected the forces of Nautius and Minucius as punishment for their lack of success, and he demoted Minucius.

Cincinnatus returned to Rome as a great hero, and might have become emperor, but he wanted nothing of this and resigned his post a mere 15 days after having accepted it. He returned to working his farm, which spanned a princely three acres, and he took no further part in the military or in politics. Through the subsequent history of Rome, his name resonated down the centuries as a synonym for valour, virtue and the complete absence of self-interest.

Such heroes aside, there is no necessary connection between charisma and virtue (or its lack). Former American President, Jimmy Carter, was a good man, perhaps too good for politics, but he lacked charisma. Perhaps virtuous charismatic personalities are really just a subset of the much larger group of good people in general. They are the ones whose narcissistic disturbances have given them the talent for leadership, and with that the potential for great power and corruption, yet despite this they choose the good. So the question really is: Why do some narcissistically disturbed persons feel compelled to work out their disturbances in ways that society deems good?

There are several preliminary dimensions to acknowledge including social influence, role models, formative environment, peer support, quality relationships and so forth. These need not be discussed here, but it can be appreciated that someone like Churchill, despite his early life difficulties, would perhaps be at least a little more likely than most to develop prosocial values and a feeling of obligation towards lesser mortals (even though many others of his class did not). He was, after all, raised in a basically benign milieu with a loving nanny, a rich environment, a supportive network of extended family and peers, a positive sense of duty and all the privileges of education, wealth and class. He never knew the dehumanising ugliness of survival values. For Churchill, societal demands were a lifeline, a way of connecting symbolically with parents who were perceived as loving and moral, and internalising them, thus making a connection in adulthood that may never have actually existed in childhood. As a child he believed that his parents were especially good, great and important, and in his childish mind this excused their neglect of him. Later as an adult, doing good deeds became his way of joining with them by living within their values, psychically connecting with them and belonging to an idealised family. Being unable to join with them physically as a child, he nevertheless joined with them emotionally as an adult by emulating them (or rather, emulating his virtuous image of them), and doing good things. Placing his own great deeds alongside theirs became a secret act of fealty to all he needed to believe they stood for. That they actually did not stand for this was beyond his ability to recognise, for were he to do so he would not have been able to avoid the conclusion that they did not really love him (in any meaningful way). Becoming great and virtuous became his way of maintaining his belief that his parents loved him.

The picture is further complicated by the fact that there are qualities and degrees of goodness that vary in time and place. The goodness that one aspires to may be impossible in a particular milieu, or given one's limitations. Mother Teresa is viewed by many as saintly, and certainly she was sincere, devout and did good works, but her ideal of goodness derived from a quite primitive, fatalistic and punitive variant of Catholicism, with little appeal to more humane spirits (Hitchens, 1977).

What is needed to explain all of this is some theory of how and why humans, both charismatic and noncharismatic, choose the good. The theory needs to be fairly independent of context, for while it can never be completely independent, if it relies primarily on situational influences then it merely explains why people in particular situations behave virtuously. What it must tell us is what it is about an individual (rather than merely the various external forces currently acting upon him or her) that inclines them toward the good, even in adverse circumstances.

To begin, pathology is morally neutral. There is no intrinsic reason why a narcissistic disturbance should place one at odds with society. The reason it so often does is because of the demands that society places upon us, demands that disturbed individuals find harder to cope with than do healthy persons. But, as Theodore Millon has advised, 'There are narcissists with good superego development' (Millon et al., 2004, p. 350).

To address this issue, Heinz Kohut put forward his formulation of Cosmic Narcissism, a state of ultimate goodness. It seems that in certain circumstances, if a person is sufficiently secure to forego their neurotic fears and desires and their need for gross narcissistic supply, they may eventually attain a quite extraordinary psychological configuration. This is a state of supra-individual participation in the world, an expanded sense of self that is dispersed through the world and that feels at one with it. Such a person merges with others, their culture or their nation, in a spirit of love, internalising them within their self, investing their self in them, and identifying with them until — psychologically at least — they experience no difference between their self and the beloved other(s). There is a sense in which the personal self dissolves and gives up its individualistic structure and cohesion, to 'expand far beyond the borders of the individual' (as the quote that heads this chapter put it), reaching a state of inter-being and absorption in the world, and of identity with it (Kohut, 1966, p. 245).

How realistic is this notion? There are some prosaic parallels. For example, much of what we routinely think of as our self is actually located externally. The person who gets hungry when the clock says 6.00 p.m. usually does not think, 'The clock says it is dinnertime so I'll eat'. Rather, they just say, 'I'm hungry,' because they are conditioned to be hungry when the clock points to 6.00 p.m. They experience the sensation of hunger coming from within, but if the clock says it is only 4.00 p.m. they may not 'feel' so hungry. Similarly, the English explorers who dressed formally for dinner in the middle of the African jungle were carrying a part of their culture with them into the African interior, and enjoying the familiar sense of home comfort and culture (or even buttressing their sanity) that came from doing so. We recognise in these examples people identifying their selves with externals. So, in principle at least, something such as what Kohut suggests is feasible; one might conceivably identify oneself with aspects of the external world.

With Cosmic Narcissism, it is as if the person regresses to their original psychological universe, the state of primordial oneness with the mother, now embodied as the community, tribe or nation. In that original fusion state, the mother was experienced as part of the self;

her body was felt as part of the child's own body, and she was perceived as infinitely good. Now as an adult the good person is able — at least partially — to abandon their neurotic fears and desires, their restricting dogmas and repetitive habits, their limiting personal perspective with its demands for ego gratification and narcissistic supply, and to live their life in a greater service to their community, culture or nation, as if it were the original mother. There is a renunciation of the ego and the id, and a relinquishing of sustaining illusions. One lives intuitively in the belief that love will conquer all, re-enacting that early sense of transcendent fusion with the total Good, now identified with both self and other in the form of one's community or some other shared collective identity, and experienced as a continuous oneness (Mann, 2000).

This may seem to be a bit far-fetched at first, but such self-transcendence may not be as remarkable as it appears. Again there are some prosaic parallels. It is not uncommon for a person to lose their self in something external, say, their lover or their child, or even in their work. Artists have this experience often, and less creative types are quite familiar with those moments of ego loss when they lose themselves in an ideal, or just in their garden. Such moments are felt to be enriching, and we like to return to them in order to escape what Alan Wheelis has called 'the prison of the self' (Wheelis, 1999, p. 17). The craftsman may lose his ego in the challenge of his materials, the nurse or teacher in the empathic demands of her charges, and even, as Leo Pirsig put it in his book *Zen and the Art of Motorcycle Maintenance*, the mechanic may lose himself in the warm purr of a beautiful machine (Pirsig, 2006).

At such times we may feel as though we participate in something greater than our personal selves, as if we belong to or are connected with something greater, something divine perhaps, or at least the heart-throb of life, in what another theorist has called 'flow' (Csikszentmihalyi, 1991). Religious people have a whole discourse of ego renunciation that is quite similar to this, and which is surprisingly consistent across all of the major religions. There are several passages in the New Testament that indicate that Jesus attained something like this state. If contemporary religious leaders or 'prophets' are anything to go by, then the trait is also fairly common among other religious visionaries. Hence other great religious (and secular) prophets may also have attained it. In *Prophetic Charisma* I reported the following comments made by one of the leaders studied, a man who led 200 followers and children in a thriving fundamentalist Christian commune. I had asked him what personal satisfaction he had gained from being a leader, and he replied:

> This is something you people find hard to understand — that people aren't doing something for the self-satisfaction and the pride that's in it and the reward of accomplishment ... cause that's the general thing that happens in society. But to me, I've got a job to do, and I know it's my job to do it, and I would have a fear of failing to do that job ... It's a tremendous responsibility. There are all these people, their lives, where they're going, and their children, that's the thing that's before you all the time. It's the thing that's before me now, night and day ... When I gave my life over to Christ when I was 21, I knew that I was gonna have to suffer a lot of things, I was gonna have to sacrifice things, and I didn't do it for some gain. I did it out of appreciation. I was so thankful to God. I was thankful for what God had done now in my heart, and in my life, the peace of mind and soul he'd given me. I was just so thankful that I wanted to do anything that would please Him, no matter what it was. And that was my pleasure ... But I'm not, I don't find it in my heart that I have much time to sit down and start thinking, say, 'Well now, isn't that great, look at what we've done'. I just see that there's so much that we haven't done and there's so little time left ... My fear has always been that I'd fail. And I wouldn't want to fail God. I'd rather die first. (Oakes, 1997, pp. 81–82)

The main points of relevance in this quote include, first, that the speaker placed the welfare of others before his own, or at least he equated the two. Second, the speaker was oriented to God's work rather than to some vanity of his own. Third, the speaker specifically rejected any personal satisfaction from doing what he did (albeit he may still have felt this at times), but rather, he was driven by a feeling of gratitude towards his God and a fear of failure. There is also a shifting aside or diminution of the personal ego in the quote, and a measuring of his work against an absolute standard ('so much more that we haven't done'). Of course, one can qualify what is said, challenging the speaker's authority to know what God wants and arguing that he probably does not know what his unconscious motives really are, and so on. But the two main psychological points to be emphasised are that the speaker seems to be as genuinely concerned for others as he is for himself (identification with them), and that he acknowledges and reveres an external criterion for action, rather than some doctrine of his own fancy; hence there is a virtuous transcendence of his ego.

The quote is reminiscent of another already presented in chapter 7 where one leader told me: 'I've become the message now' (Oakes, 1997, p. 13). It is also similar to St Paul when he wrote, 'It is no longer I who live, but Christ who lives in me' (Galatians 2:20). This absorption of the individual self into a greater, almost metaphysical sense of being was reported by several leaders in my original study, and from both religious and nonreligious orientations, typically:

> I am a function. I just live my function ... See, at this point there's only the function, there's only function, there's only who I am expressing ... In the past I was looking for acknowledgment or trying to get something. Now to me all that is a joke ... (Oakes, 1997, p. 84)

Prophetic Charisma described two kinds of charismatic prophets. The first or pure charismatic was truly self-enclosed, self-contained and self-referential in his thinking. This corresponded to what Erich Fromm has called malignant narcissism, in which the leader's sense of virtue comes not from what he does or produces or achieves, but from something he has or is, or some inner quality. Hence, he does not need to be related to reality in any way; indeed, it is an advantage not to be because this enables him to inflate his grandiosity to vast proportions in order to avoid discovering that his inspiration is merely a product of his imagination. Thus, malignant narcissism lacks the self-corrective element that arises from interaction with reality; it is crudely solipsistic and xenophobic (Fromm, 1964, p. 77).

Benign narcissism, on the other hand, while also a product of pathological early life processes, nevertheless involves identification with an external frame of reference in the form of a system of transcendent values that is related to maternal virtue. Hence, this form of narcissism is self-checking; to do God's work the leader must be related to reality in some positive way and this curbs his narcissism and keeps it within prosocial bounds.[1]

Quite how this comes about is complex; Kohut at one point invoked genetics, while elsewhere he said that it came about through psychological mechanisms.[2] The most important distinction is a practical one, concerning whether or not the developing narcissist aligns himself with an external framework of values that he embraces to guide his behaviour. Does he subordinate himself to a transcendent ethic that he chooses to live 'up' to, some external standard that he embraces; or does he locate his god or ultimate authority 'within', and thus come to lead others by whim and opportunistic fiat? Does he follow the accumulated wisdom of his community, or does he intuit his course through life by purely subjective personal dictates, regardless of the needs of the people he leads, and without recognising his responsibility to them? In sum, it is about whether he chooses to be pulled by his ideals or pushed by his ambitions. If he does the former, then he is attempting to embrace the good life, as his culture has formulated it; whereas if he does the latter, he is setting himself up as the source of ultimate good for those he leads. They can know no greater authority than he, for he is the way, the truth and the life.

In *Prophetic Charisma* I listed 15 distinctions that may be made between the two types, many of them being fairly technical and of interest only to students of religion. But this list also included some behavioural correlates, one of them being that malignant leaders tended to indulge in illegal behaviours more frequently than did benign leaders. Malignant leaders also retained power within themselves for as long as they could, whereas benign leaders accepted a gradual decline in their power and even planned for succession. Furthermore, benign leaders tended to be utterly consistent, whereas malignant leaders were arbitrary and consistent only in their inconsistency (Oakes, 1997, pp. 183–184). The acceptance of an external authority is the key; the charismatic leader either looks upward to something transcendent, saying, 'You are perfect and I am part of you', or he looks defiantly inward and declares, 'I and the Father are One'.

Hence the processes involved in diminishing one's own ego and identifying one's self with suffering humanity, God, or the 'Good', involve consistent psychological actions of a basically narcissistic nature. There is a self-love that is so expansive and self-transcendent that it evolves into identification with, and love for, humanity as a whole, still felt as oneself but now in a greatly enlarged sense. The psychological operations that produce such individuals may be quite ordinary and available to us all, and are perhaps no more profound than the way one loses one's self in one's gardening, or relies on the clock to get hungry. But the readiness for this, and the desire to do so, may have taken a lifetime of preparation.

It may even be, as Nelson Mandela suggested — in a classic statement of cosmic narcissism during his 1994 inaugural speech upon becoming President of South Africa — that we have things the wrong way round. Perhaps cosmic narcissism is a more natural, holistic state than the fragmented, separated, ego-driven existences that we routinely lead. Perhaps it is the most natural state of all, something that was taken from us very early by unfortunate life circumstances, yet which a few retain and all may seek. Mandela told his listeners:

> Our deepest fear is not that we are inadequate. Our deepest fear is that we are powerful beyond measure. It is our light, not our darkness, that frightens us. We ask ourselves 'Who am I to be brilliant, gorgeous, talented, and fabulous?' Actually, who are you not to be? You are a child of God. Your playing small doesn't serve the world. There's nothing enlightening about shrinking so that other people won't feel insecure around you. We were born to make manifest the glory of God that is within us. It's not just in some of us, it's in everyone. And as we let our own light shine we unconsciously give other people permission to do the same. As we are liberated from our own fear, our presence automatically liberates others. (Mandela, 1994)

Mandela's early life is instructive in this regard (Guiloineau, 2002). He was descended from tribal kings and his father functioned as Prime Minister among the Thembu people. His father had several sons by four wives, but by the time Nelson was born the other sons had already left home. As he described it, his was an especially close-knit family. 'I had mothers who were very supportive of me and regarded me as their son ... not as their stepson or half-son as you would say in the culture amongst whites. They were mothers in the proper sense of the word' (Sampson, 1999, p. 5). However, he seems not to have been close to his father, a nobleman who, in Nelson's infancy, was dispossessed by a white magistrate for 'insubordination'.

The circumstances of this dispossession are suggestive. The initial cause was a complaint lodged against his father by someone Mandela describes as 'one of my father's subjects' (Mandela, 1994, p. 7). When the father refused to appear before the magistrate on summons he was immediately deposed by this official. Mandela writes that although he was merely an infant at the time and thus he knew nothing of this, he was not unaffected. His father lost his fortune and his title, and he was deprived of most of his herd and lands and the revenue that came from these. Because of this sudden impoverishment his mother had to move to another village where she had the support of friends and relations.

This can only have been a catastrophe for the family, causing anxiety and distress to everyone, and affecting the emotional climate in which Nelson was raised. It happened in his first year of life, and doubtless it would have preoccupied his father for years after. With this disaster at the forefront of his mind, and with four other families to care for, the old man must have been quite unavailable to his son and probably remote when he was present. He moved between the homes of each of his four wives, homes that were separated by many miles, and he died when Nelson was nine years old. Therefore, they probably were never very close and never got to know each other deeply; his father had visited him and his mother only monthly. Nevertheless, Nelson recalled his childhood as very happy. No doubt village life with four mothers would have been happy for him, but the absence of a fatherly relationship and the undercurrent of injustice and impoverishment must have left its mark on one who had been the undisputed darling of his mother(s), and who later went on to fight injustice in the most spectacular way.

Again there are hints of a dual developmental path. Objectively there was tremendous loss, a plunge into poverty, insecurity and stress, and quite probably some fragmentation of the family nexus. Yet against this background young Nelson was greatly elevated; told he was 'special' (Lodge, 2006, p. 2), and perhaps even treated as the future restorer of his father's legacy, the hope of things to come. We cannot know for sure what were the subjective meanings that he ascribed to these external events, or even the degree to which he was aware of them, however distantly. Nor can we know how he might have felt about the disturbed emotions and behaviours that would have filtered down to him from the other members in his family. But most likely there would have been some feeling of deprivation, anxiety and rage, alongside his sense of specialness.

Mandela has said that after his father's death he felt, 'not so much grief as cut adrift' (Lodge, 2006, p. 16). For although only nine, he was now separated from his mother(s) and moved to the 'Great Place' to live with the Thembu chiefs. Again there is a hint of a dual developmental pathway; he bore his losses and learned to repress his emotions, while being groomed like his father before him to be a counsellor to the king. His background of feeling special amid his family's losses, and his newly elevated status amid his own emotional losses, must have had a great effect on him. Quite how is unknown, but he became remote and self-contained. He was privileged, but at a price; in his autobiography there is a recurring, poignant image of a child attempting to fill the clothes of a father: Nelson after his father's death, and his own son after Nelson's divorce.

Nelson was the first of his family to attend school, eventually graduating from university as a lawyer able to quote Shakespeare, and influenced by Churchill and Gandhi. One of his early role models was General Kutuzov in Tolstoy's War and Peace:

> Kutuzov defeated Napoleon precisely because he was not swayed by the ephemeral and superficial values of the court, and made his decisions on a visceral understanding of his men and his people ... to truly lead one's people one must also truly know them. (Mandela, 1994, p. 585)

In time Mandela emerged as a leader among the African National Congress (ANC), the largest political party opposing apartheid in South Africa. This led to him founding the Umkhonto wing, a quasi-military style group within the ANC, and becoming a saboteur (he was never a terrorist or a communist). He was arrested and spent a total of 27 years in prison. Although he was not brutalised in prison, conditions were harsh; however, he was allowed to study and he had access to his fellows in the movement. This experience seems to have transformed him. A psychological report undertaken in 1981 — by which time he had been in prison about 20 years — described him as 'exceptionally motivated', with 'outstanding personal relations', who was pragmatic rather than ideological, not racially embittered, self-critical and reflective, and possessing an unbelievable memory. As biographer Tom Lodge has described it, 'Imprisonment had served to increase his 'psycho-political posture' and enabled him to acquire an especial charisma' (Lodge, 2006, p. 150). In sum, he endured, and when in 1990 the political situation had changed, he was released from jail. He was elected the first Prime Minister of the newly demo-cratic Republic of South Africa in 1994.

With hindsight it is now clear that the African National Congress leaders — Mandela, Oliver Tambo, Walter Sisulu and others — were the cream of their people, standing head and shoul-ders intellectually above the more limited and rigid Afrikaner politicians they dealt with (Botha, Vorster, Malan and their ilk). Mandela even suggests that both Tambo and Sisulu were actually smarter than he, and in addition Sisulu was highly confident. At first Mandela was something of a 'show-boy' among them (Lodge, 2006, p. 35), and he slowly groped his way towards his own philosophy through their tutelage. They were all inspiring speakers, but what eventually distinguished Mandela was his temperament, the centred-ness of his being. He seemed to know what was right, more reliably, accurately and frequently than did the others, possessing a dominat-ing presence and 'an inexplicable serenity' (Lodge, 2006, p. 11). He appeared to be more benignly substantive. He had the kind of con-gruence that comes from knowing one's own heart and being grounded in what is true and good. It may also have given him balance knowing that the architects of his family's impoverishment came from both White and Black races, for he has been extraordinarily generous and consistent in his dealings with all ethnic groups. He also had known decent and friendly Whites since his childhood (Lodge, 2006, p. 11).

Is Mandela a charismatic personality or merely a strong, benign leader? In contrast to leaders such as Churchill, Freud and Hitler, with their obvious and abundant neuroses, he appears to be the very model of good psychological health. However, there are sufficient idiosyncrasies to his personality, especially hints of a 'false self' that nevertheless serve a higher (political) truth, to enable him to be categorised as charismatic, perhaps even a consummation of charismatic personality development.

For example, by his own admission Mandela has no close friends. He guards against emotional friendships and has developed a 'total politicisation of being', an observer adding that, 'It was a price I wouldn't like to pay, but it gave him a remarkable integrity in political life' (Sampson, 1999, p. 499). He is similarly defended against depression, saying in an interview that he has 'never had a single moment of depression' (Sampson, 1999, p. 230) because he is so convinced of the rightness of his cause and its ultimate success. Although he admits to feeling weak and needy when alone, he comes alive in company, seeming to live through others. His former secretary Mary Mxadana has recalled that, '... in company he lit up with every visitor, but by himself he would suddenly look exhausted, as his welcoming smile turned into a grim circumflex. Which mouth should the artist show?' (Sampson, 1999, pp. 498–499).

The theme of his ever-changing personae recurs in all the biographies of him. He is a master of impression management and he has a deep understanding of the symbolism that he needs to embody as a healing leader. He carefully manages his public displays of emotion, and has an exceptional ability to make whoever he is with feel special (Lodge, 2006, p. 213). So extraordinary and complete is his ability to be the person that others need him to be, that one historian has described his effect on Afrikaners as producing 'a state of charismatic bewilderment' (Lodge 2006, p. 213). Richard Stengel, the American journalist who collaborated with him on his autobiography, has commented, 'The man and the mask were one'. Politicians and diplomats waited for the mask to slip but it never did, because his personal feelings were subsumed by his political life (Sampson, 1999, p. 498). He felt offences to his dignity far more deeply than physical wounds or political insults (Sampson, 1999, p. 511). At the end of the final chapter of his authorised biography on Mandela, author Anthony Sampson focuses on this fusion of image and reality that has occurred in Mandela, writing: 'Mandela has remained a master of political images, but they have become part of his own personality and history … his biography in the end converged with his mythology … ' (Sampson, 1999, p. 585). He became his function.

Mandela is probably the most respected man on the planet, but he concedes that he has made mistakes; he failed to deal effectively with the AIDS crisis, and his economic management also remains controversial. But perhaps what this really shows is simply that, by the time a charismatic leader is installed, the problems he has been mandated to deal with are likely to be very severe indeed (nothing could have prepared him to for the AIDS crisis). It also points to another difference between benign and malignant charismatic personalities; cosmic narcissists are able to admit mistakes, whereas most narcissists are not. This is because their acceptance of a transcendent ethic of responsibility enables them to look somewhat objectively at their own behaviour. The injunction frequently encountered in corporate culture to 'never apologise and never admit a mistake', is a symptom of more primitive psychological make-up.

Mandela has also paid a high personal cost in terms of the breakdown of his first two marriages and the loss of his relationship with his son, who became alienated from him in his last years and who died while he was in prison. While he has a certain vanity — he enjoys the attention of beautiful women, intellectuals, the talented and the rich and famous (and he loves his clothes), all this seems not to have corrupted him. He is sustained by wiser passions. Those who meet him say he possesses a natural dignity, but this sanitises his image; a more apt descriptor is 'majesty'.

Mohandas Gandhi provides another illustration of the life of the benign charismatic personality.[3] Like Mandela, he too was born into the aristocracy. His father was the Prime Minister of a small Indian state and his mother was a saintly zealot. His home life was cultured and privileged, and he was shy and bookish. Betrothed at age seven and married at thirteen, he has written that he knew what to do on his first night of marriage from the memories he had from his previous life.

To the modern mind, Gandhi's thorough and mostly unquestioning immersion in religious superstition is an odd theme. He could be critical of it, later becoming quite bitter at his parents for wedding him at such a young age, describing himself and his wife Kasturbai as 'married children', but it seems to have overwhelmed him. He was very much a child of 19th-century India.

Gandhi's transformation into a holy man occurred after some rebellious youthful (extra-marital) experiments in meat-eating, smoking, thieving and a visit to a brothel, which led him to feel so guilt-ridden that he contemplated suicide. Instead he confessed to his father, then confined to his deathbed, who uncharacteristically — and tearfully — forgave him. This experience of 'sublime forgiveness' so inspired the 16-year-old Mohandas that he held to the principle of ahimsa or loving non-violence for the rest of his life.

This lesson was greatly reinforced soon after when his father, still restricted to bed, died in circumstances that left a deep imprint on him. It was Mohandas's nightly task to give his father his medicine and to massage his legs, leaving only after the old man had fallen asleep. One night as he was giving the massage an uncle offered to relieve him. Feeling some sexual longing, Mohandas returned home quickly to awaken the pregnant Kasturbai for sex. But mere minutes into this a servant knocked at his door to say that his father had died. Subsequently, the child Kasturbai bore also died three days after birth. Thus, while still an impressionable adolescent, sex, guilt and loss became closely associated in Gandhi's mind. He retained this mindset all his life, concluding that 'morality is the basis of things and truth is the substance of all morality' (Gandhi, 1927, p. 47), and that renunciation is 'the highest form of religion' (Gandhi, 1927, p. 78).

Aged 18, Mohandas went to England to study as a lawyer. This gave him 'a long and healthy separation' from his wife (Fischer, 1951, p. 36), and led to a continuation of his 'experiments' aimed at transforming himself, mostly to do with dietary and self-management regimes. He left England as soon as he graduated, but on his return to India he performed poorly as a lawyer. An invitation to represent the oppressed Indian minority in South Africa gave him another opportunity to spend time away from his wife, so he took it. While there he found that by pursuing legal settlements out of court he could avoid the problem of his shyness. Biographer Louis Fischer says that he remade himself in South Africa, undergoing a 'second birth', discovering his talent as an organiser and an aptitude for nursing. Kasturbai joined him there and he acted as midwife in the births of his own children. Later he took a vow of celibacy.

Gandhi was now successful and influential in his own right. He had found his niche, and by age 37 he felt ready for a great mission. His subsequent return to India and his leadership of the independence struggle is history, but his motivations were mixed. Author Robert Payne has written how, on his return from South Africa, Gandhi was:

> ... searching for a single cause, a single hard-edged task to which he would devote the remaining years of his life ... he dreamed of assembling a small army of dedicated men around him, issuing stern commands and leading them to some almost unobtainable goal. (Grenier, 1983, p. 55)

In short, despite whatever noble and progressive considerations were in his mind, Gandhi was also searching for a vehicle for his own aggrandisement; the Gandhi of popular pacifist imagery came much later. In South Africa he had recruited Indians into the British army (although they only acted as stretcher-bearers), and he was later to make several promilitant statements (Fischer, 1951, p. 202; Grenier, 1983, p. 68). He had no particular concerns for Black South Africans, and when the Zulus rose up he volunteered to organise an Indian brigade to suppress them. He was even awarded the War Medal for valour under fire. Back in India during World War I he offered to recruit Indians for the British war effort.

Gandhi's pacifist ideology completely misled him when Hitler arose. His advice to the Jews was to accept martyrdom at the hands of the Nazis, effectively to commit collective suicide, thus creating a moral triumph that would be remembered for ages to come. He offered the same advice to the Czechs, later advising the British that, 'Hitler is not a bad man ...' (Grenier, 1983, p. 80). His grandiosity was so inflated that he even attempted to convert Hitler, writing him a letter (beginning 'Dear Friend ...') that urged him to embrace all mankind 'irrespective of race, colour, or creed' (Grenier, 1983, p. 81). When the Japanese threatened to invade India, his proposed strategy was to allow them to occupy as much land as they desired and then to 'make them feel unwanted', apparently imagining that their feelings would be hurt and thus they would leave (Grenier, 1983, p. 81).

In later life, Gandhi sometimes slept with naked teenage girls to 'test' his vow of chastity (apparently they fought 'hysterically' for the privilege), and he routinely administered enemas to them and they to him (Grenier, 1983, p. 42). This latter was part of his lifelong preoccupation with diet, health and hygiene, which in turn reflected the traditional Hindu fetish for purification and pollution. He actually knew little about health, but he considered himself an expert and he became fanatical, and even at times emotionally unstable about it. He once had a confrontation with Margaret Sanger over birth control (she was an advocate while he was a fanatical opponent of sex for pleasure) that was so forceful that he later had a nervous breakdown (Grenier, 1983, p. 41). At his ashram he dictated every aspect of daily life and every morsel of food to his followers (Grenier, 1983, pp. 45–46). When Kasturbai lay dying of pneumonia and British doctors insisted that an injection of penicillin would save her, Gandhi refused to let this alien medicine be given to her and she died, but later when he contracted malaria he accepted the equally alien quinine, and later still an appendectomy (Grenier, 1983, p. 34).

Gandhi's own wisdom as a father was questionable. He refused to allow his sons to be educated, he disowned his older son Harilal for wishing to marry, and he banished his younger son for giving the struggling Harilal a little money. In due course, Harilal became estranged from him, converted to Islam, attacked him in print, indulged in women and drink, and died an alcoholic.

Gandhi was so erratic and unpredictable that he was thought by some Indians, including V.S. Naipaul, to have possibly delayed Indian independence for many years. He was difficult to work with, autocratic and intolerant of opinions other than his own. His proposed political reforms were moderate; although he wanted greater tolerance for the Untouchables, renaming them 'Harijans' or 'children of God'. He stuck with the basic Varna caste system until the end of his life, believing that a man's position and occupation should basically be fixed at birth. He claimed to long for a personal guru, but found that none of those he met were able to live up to his high standards for the ethical and spiritual life, though this may merely reflect his reluctance to accept another's authority. His final statement on the matter was that one gets the guru one deserves (Gandhi, 1927, p. 93).

Gandhi illustrates the mixture of egoism and altruism, pathology, superstition and spirituality that often intermingles in the benign charismatic personality. Despite his modern education he could not transcend his Hindu roots, and for all his strategic vision he was not able to turn it inwards and analyse himself, to identify the ways in which some of the attitudes he opposed were also present within himself. Yet for all this, his basic goodness is indisputable, as is his place in history. His preparedness to put himself personally on the line, his teaching by example and instruction of nonviolence, his endless hard work in organising and debating, his tactical abilities and his use of fasting and passive resistance as political tools, place him at the front rank of freedom activists. He was not the first to advocate civil disobedience or non-violent political tactics; Henry Thoreau was an important precursor and there have been others. But Gandhi was the first to make non-violence work on a large scale, and he directly inspired Martin Luther King, Nelson Mandela and others. (His wonderful sense of humour has also sometimes been overlooked; for example, when asked what he thought of English civilisation he replied that he thought it was an excellent idea.)

Gandhi may well have stood in the way of more rational voices working for Indian independence, but none of these others embodied the Indian mentality as he did. He expressed both the highs and lows of his culture; the ahimsa doctrine applied to large-scale politics, and the regressive authoritarianism, superstition and pollution fetishism of traditional Hinduism. The honorific given to him, 'Mahatma', literally: 'great soul', and the other descriptor most frequently used for him, simply 'baba' or father, suggests the combination of the spiritual and the familial that he symbolised — the 'Great Father of India' (Wolpert, 2001). Perhaps more than any of the other voices of his time, Gandhi understood that regardless of his own personal limitations, at that particular time and place history inclined more to the strong than the wise. The bloodbath that occurred after the British partitioned India can hardly be blamed on him; subsequent events suggest that there was something very dark lurking in Indian culture.

To generalise, the good leader retains a close emotional connection with the primitive state of primary narcissism while also embracing the transcendent ethical stance of his culture, despite whatever other disturbances afflict his or her personality, and regardless of whatever other contextual factors limit him or her. Cosmic narcissism, the state of undifferentiated goodness, may be to the individuated or self-actualised 70-year-old what primary narcissism is to the newborn, a recapitulation of themes of perfection, goodness, beauty and love, but now on a communal or even a planetary scale rather than on an infantile scale. The ordinary person who chooses the good, and the charismatic personality who does likewise, may be drawing on the same resource of primitive union with love and perfection, a union that knows no distinction between self and other. It may be frightening to us because as adults we recognise that such a state is extremely vulnerable, or at least that it was once to the infant. Courage is required to transcend the many painful and inhibiting experiences that have made us what we are, and to choose to 'regress' to (or really to reclaim) such a primitive birthright. Nevertheless, perhaps anyone who lives long enough and well enough may eventually return to their emotional roots, wherein the psychological meaning of heaven is mother-love.

Several points arising from this theory deserve further clarification. First, it is not assumed that the person who develops cosmic narcissism does so continuously. We may all glimpse moments of it at times, but we may also fall far short of it often. Churchill, for example, sometimes behaved badly, and as with Mother Teresa, his notion of goodness was quite limited, in his case, to his rather 19th-century Tory sensibility. Society curbs both our antisocial impulses and our tendency towards narcissism, but a consistent lifetime pattern of choosing virtue, especially when one is able to transcend social controls, as charismatic personalities often are, has to originate somewhere, and probably this is in early childhood.

Second, it is not assumed that only the most psychologically healthy individuals evolve cosmic narcissism. The pathways to it may be many, varied and complex. The example of Churchill suggests that cosmic narcissism may co-exist with other neurotic disturbances. It is also no accident that most of the great exemplars of cosmic narcissism were culturally and socially well-placed to do so. Churchill was the grandson of a duke, Mandela was of the royal Thembu bloodline, and Gandhi also was born into the aristocracy. We know nothing of Cincinnatus, but as a Roman senator and a landowner he too probably had aristocratic origins. George Washington came from minor Virginian gentry, and the fable of little George being unable to lie about chopping down the apple tree catches the flavour of someone solidly bonded to an external ethic while still a child.

What seems to happen is that infantile narcissistic disturbance gets combined with very ethical social influences at a later but still quite early stage of life (the formative years of childhood), in such a way that the resulting life direction combines a narcissistic mindset with deeply moral convictions. Why this happens with some children and not others remains unknown, but it is not a mystery. What may be significant is whether or not the original deprivation, neglect or abuse was personally directed or impersonal. Freud and Churchill suffered neglect and trauma as children, but for the most part it was not personally directed at them, or only incidentally so (in the case of Churchill's father's disdain), whereas the kind of abuse suffered by Hitler at the hands of his father was personal, of quite a different magnitude. Presumably there are also other subtle environmental and dispositional influences.

But perhaps the main influence comes from the father. If the primary carer (usually the mother) has the most influence on the structure of the developing child's mind, determining the nature of the child's relationship with itself (narcissism or realism), then it is likely to be the secondary carer (usually the father) who most influences the child's relationship with the world (benign or malignant), for he is the family agent most actively involved with the world.[4] His example of virtue and prosocial engagement is likely to become the model for the child's own engagement. Of course, the father's example may be symbolic rather than actual, as it was with Churchill's father. The child's perception of his father may also be illusory or compensatory, as it clearly was in Churchill's case. It may even be mediated be a third party, in Churchill's case his nanny, the benign Mrs Everest, who seems to have been his earliest bridge to the external world. There is a need in the developing child's mind for some model of social participation, and the example it receives primarily determines its subsequent behaviour.

Third, what is defined as good varies according to time and place, and actions deemed good by one group may not be so evaluated by another group. For example, many Polish people still loathe Churchill to this day for, as they see it, betraying them; and Washington only released his slaves on his deathbed (although he was the only Virginian founding father to actually do so). The desire to do good does not automatically bestow on one an eternally valid idea of what the good actually is, and cosmic narcissism is usually limited to one's own community, culture or nation. Probably only rarely does it embrace the entire world, as perhaps it did with Socrates, Buddha and Jesus. Even when it does, this might not be an entirely good thing, as Gandhi's advice to the Jews indicates.

Finally, to a modern charismatic saint, George Baker, a short, African–American preacher who came to be known as Father Divine. Baker was one of the most remarkable religious leaders ever. In the classic manner of the charismatic leader who arises to inspire an oppressed people, Father Divine appeared at a difficult time for African–Americans. In the decades before the freedom marches, the few role models they had were sportsmen and jazz musicians. Father Divine added to the list by embodying incorruptible virtue, and by providing jobs and caring for the poor years before social security. He became a refuge and a hope, and for some, God Almighty! Baker was born about 1864, probably of slave parents. Little is known about his early life, but he apparently bucked racism as a youth and was in trouble several times. He worked in menial jobs and trained as a preacher. In about 1900 he teamed up with two other black preachers and took the name 'The Messenger' (the others called themselves Father Jehovah and St John the Vine), and together they travelled as a flamboyant preaching team.

In 1912 the trio broke up, and three years later The Messenger arrived in New York. He settled in Brooklyn with a few friends and studied the newly founded church run by his old comrade St John the Vine. St John taught that God lived in every person (not in heaven), and this implied that in some sense everyone was God. St John certainly claimed to be. This led to many of his followers sporting mock gold crowns and purple robes, and taking names such as Father Obey, Elizabeth of the Fiery Chariot, and Saint Paul, to express their divinity. The Messenger learned from this, and when he founded his own church he permitted only one God in his cosmology — Himself — whose rule was unquestionable.

Next, The Messenger started an employment agency providing domestics and unskilled workers who were members of his church, hence honest and reliable. They gave him their earnings in return for lodgings and food. They were expected to live a celibate life and to take their example from The Messenger and his wife Peninah, whose marriage was legal and spiritual but not sexual. Throughout his career, Father Divine taught that sex was sinful.

After several more name changes, The Messenger became Father Divine in about 1916, and named his movement the Peace Mission. He elaborated a myth of his own divinity; to queries about his birth he would reply, 'I wasn't born, I was combusted one day'; or, 'Before Abraham was, I am'. He combined this mystique with a genuine worldliness; he worked in the garden of his headquarters on Long Island and became a competent business administrator and investor. In time he became spectacularly successful; on Sundays hundreds of his followers flocked to his services to feast and listen to his choirs, the Rosebuds — young girls, the Lilybuds — adult women, and the Crusaders — males.

However, Father Divine became too big, at least for a Black on Long Island in the 1920s. The police investigated, and after they found that they could not prosecute him for anything much, they sent in an undercover female agent hoping to compromise him. She attempted to seduce him but failed, and soon after she became a follower. After this the police, under pressure from nearby residents who objected to the large gatherings at his Sunday services, raided the headquarters and arrested Divine for disturbing the peace. He was arraigned before Judge Lewis Smith in December 1931. Smith was obviously hostile to Divine, cancelling his bail and remanding him to prison throughout the trial. After Divine was found guilty, but with a recommendation by the jurors for clemency, Smith brought down the stiffest sentence possible: one year in prison and a $500 fine.

Father Divine went with dignity to his prison cell. Three days later Judge Smith, aged 50 and in prior good health, died of a heart attack. When Divine was told of the death he replied ruefully, 'I hated to do it'. A few days later he was released from prison on appeal, and he was subsequently acquitted because of prejudicial comments that Smith had made during the trial.

The prestige and fame of Father Divine and his group rose enormously. The organisation subsequently went from strength to strength, establishing a profitable hotel business with operations in other large cities including Philadelphia, Newark and Jersey City, and expanding its employment agencies.

A central part of Father Divine's teaching was self-respect. His followers were told to give their employers an honest day's work, to refuse tips and to find satisfaction in their jobs. Swelling with pride, many changed their names to such exquisite vanities as Miss Sunshine Bright, Mr Brilliant Victory, Miss Magnetic Love, Miss True Sincerity, and Mr Onward Universe. None of the followers were permitted to receive welfare. At its height in the 1940s, the Peace Mission probably had about 100,000 members, of whom about 90% were Black, with females outnumbering males three-to-one.

A few years after Peninah died in 1940, Father Divine took a second wife, a young White follower known as Sweet Angel, his 'spotless virgin bride'. At the time interracial marriage was illegal in 30 states, so the wedding was something of an event. The banquet was probably the most lavish ever given in America; there were 60 different kinds of meat, 54 vegetables, 20 relishes, 42 hors d'ouevres, 21 breads, 18 drinks, 23 salads and 38 desserts; in all, 350 different foods were served over 7 hours.

There are many legends about Father Divine. He was the subject of a racist vendetta launched by the Hearst newspapers, and was consequently weakened by scandals. However, he retained his spirit to the end, dying in 1965, allegedly 101 years old (it is now known that he was born in 1876). Since his death, the Peace Mission has declined to the point where today it has only a few members; this is a typical outcome for such celibate groups, including the Shakers. But in his life Father Divine brought pride, hope, dignity and wellbeing to many people excluded from mainstream society. In a scholarly book on unusual groups in America, sociologist William Kephart described Father Divine as 'a man of infinite goodness' (Burnham, 1979; Kephart, 1982; Watts, 1995).

To summarise, in order to understand virtuous leadership we must look beyond the myth of the saint. The great saviour, avatar or messiah who combines the virtues of wisdom and strength, courage and goodness, all to an exceptional degree, is probably a figure of myth and faith. Unlike the saint, who usually operates within a fairly circumscribed context, and is seldom called upon to balance the ends with the means of human conflict, the leader operates fully in the real world. He usually does so without the benefit of exceptional wisdom or intelligence. Invariably he brings along his baggage of neuroses, conceits, illusions, fallibility and other human frailties. Hence, we ought not to expect him to possess some moral compass that we ourselves lack, or to make better choices than we ourselves might. That he gets the job done without causing too much damage is good enough.

But if it turns out that the leader actually does have some virtuous moral compass that others lack, enabling him to make better choices than they might, as some leaders do, then this is an important question to address. How do they do this? It seems that they do so by accessing an archaic state of primary narcissism or love, and by following the example of some previously felt and internalised, benign and hopeful relationship with the world, which they have grown to identify with their own being. In this way, over a lifetime, mere narcissism may beget moral capital (Lodge, 2006, p. 224).

Endnotes

1　Fromm's distinction between the benign and malignant variants of narcissism corresponds approximately to similar distinctions made by Kohut between messianic and charismatic personalities, and also by Weber between ethical and exemplary prophets (Oakes, 1977, p. 41).

2　The relevant quotes from Kohut are:

> A part of the child's narcissism is transferred upon his superego. (Kohut, 1966, p. 245)

> Whether [narcissism] contributes to health or disease, to the success of the individual or to his downfall, depends on the degree of its deinstinctualization and the extent of its integration into the realistic purposes of the ego. Take for instance Freud's statement that 'a man who has been the indisputable favourite of his mother keeps for life that feeling of a conqueror, that confidence of success that often induces real success'. Here Freud obviously speaks about the results of adaptively valuable narcissistic fantasies which provide lasting support to the personality. It is evident that in these instances the early narcissistic fantasies of power and greatness had not been opposed by sudden premature experiences of traumatic disappointment but had been gradually integrated into the ego's reality-oriented organisation. (Kohut, 1966, p. 253)

> Just as the child's primary empathy with the mother is the precursor of the adult's ability to be empathic, so his primary identity with her must be considered as the precursor of an expansion of the self, late in life, when the finiteness of individual existence is acknowledged ... The achievement—as the certainty of eventual death is fully realized—of a shift of the narcissistic cathexes from the self to a concept of participation in a supra-individual and timeless existence, must be regarded as genetically predetermined by the child's primary identity with the mother ... the genuine shift of the cathexes toward a cosmic narcissism is the enduring creative result of the steadfast activities of an autonomous ego. (Kohut, 1966, p. 266)

> Such a 'toned down,' less grandiose and more attainable ego ideal permits the normal narcissistic gratification of living up to the internalized ideal parental images, and this gratification in turn reinforces self-esteem, one's confidence in one's own goodness and one's trust in gratifying object relationships. (Kernberg, 1975, p. 240)

3   See especially: Erikson (1969), Fischer (1951), Gandhi (1927), Grenier (1983), and Wolpert (2001).

4   Again, it must be stressed that no unwarranted assumptions or prescriptions about women or men are intended here. What are being referred to here are the functions that are most commonly associated with men and women, mothers and fathers. Of course, in the vast array of social variety there will be inevitable exceptions in which the primary carer is the father while the secondary carer is the mother.

CHAPTER FOURTEEN

· · · · ·

Other Enabling Traits

Heavens how that man can talk. He has the faith. He can get himself to believe anything, anything!

Joseph Conrad (1902)

Charismatic personalities are different from others in healthy, as well as unhealthy ways. Because of their narcissism they may be stimulated to exceptional development of quite normal faculties, in much the same ways that qualities such as giftedness (Leonardo), prodigality (Mozart), and genius (Einstein) arise among noncharismatic achievers. Michael Murphy of the Esalen Institute made a study of such possibilities, and identified 12 psychological functions that he believed have 'metanormal' capabilities; that is, traits that may be developed to extraordinary levels (Murphy, 1992). The most widely studied metanormal trait — arithmetical ability — has been shown to depend upon an immensely rich repertoire of intuitive cues and computational strategies specific to each individual 'savant' who possesses it, one of whom has now written a book to demystify his exceptional ability (Tammet, 2009). Today the brains of such exceptional individuals are investigated scientifically (Phillips, 2008), as tomorrow their genes will be. This chapter considers several traits relevant to charisma, specifically social insight, memory, creativity and will, while also discussing the mindset of charismatic personalities and briefly mentioning two promising lines of investigation from contemporary brain research.

The Charismatic Mindset

The quote from Violet Bonham Carter about Winston Churchill that headed chapter 2, and which reported that his relationship with all experience seemed to lack 'the shock absorber of vicarious thought', implies that he experienced himself and his environment differently from how others do. Yet some of this difference leaked out and could be detected by astute observers. While Carter's account of Churchill is informative, a richer description of the charismatic mindset would require reports from several trained observers over time. Fortunately, such a repository exists. Some of the best descriptions of the charismatic mindset can be found in the biographies by Martin Shepard and Jack Gaines (Gaines, 1979; Shepard, 1976), of master therapist Fritz Perls, founder of the influential Gestalt therapy. Most of the informants in these books were experienced therapists, skilled in interpersonal observation and analysis.

The first thing to appreciate about Perls is that whatever he was on one occasion, he almost certainly would not be on another. Claudio Naranjo said of him, 'With Fritz, of course, you could expect anything, from the best to the worst. Perhaps his creativity required the context of such open-mindedness' (Gaines, 1979, p. 247). Another informant, Cynthia Sheldon added, 'Fritz was more inconsistent — sometimes perfectly on, sometimes not on at all. He would phase in and out' (Gaines, 1979, p. 247). Hence, there appears to be an inconsistency to charisma; it may not be constantly present.

Next, there was the detached involvement that Perls brought to (almost) every encounter that enabled him to remain differentiated from others. The term 'differentiation' is used here in a specifically psychological sense and is derived from the work of pioneering marital theorist, Murray Bowen. It is defined as the ability to securely hold onto oneself — one's own thoughts, emotions and sense of oneself — when in close with another, so as not to be unduly influenced by them (Kerr & Bowen, 1988). In this regard, Shepard has noted that there was something quite distant about Perls, an 'utter detachment' (Shepard, 1976, p. 100). Gaines similarly reports a 'remote quality' to him (Gaines, 1979, p. 309). It seems that the charismatic personality always holds something in reserve.

Closely related to his differentiation was his nonjudgmental and permissive openness. Another of Gaines's informants added that Perls 'didn't have the normal fantasy overlay' (Gaines, 1979, pp. 68–69). This description, so reminiscent of Carter's description of Churchill, suggests that he did not see others through a veil of his own needs and fears, illusions and judgments. This may explain why they trusted him and were willing to open up. They felt accepted because he really heard and understood them (he has been described as 'hearing one to death'). This enabled him to get on their wavelengths quickly; they felt reassured by the immediacy of his attention to them.

Perls also had a profound sense of his own authority; he 'simply knew what he was doing to an uncommon degree', and his 'sense of responsibility to himself was deep' (Gaines, 1979, pp. 128, 149). This sense of inner authority was not derived from any status or qualification, but from his own self-reliance through adversity. He had proven himself to himself, so he was unafraid of his weaknesses and was intensely congruent. Someone who has come through a very disturbed childhood (rather than merely having had 'issues') may develop such deep trust in themselves. As the film *Damage* (Malle, 1992) put it, 'Damaged people are more dangerous than others because they know they will survive'. If someone is afraid that a particular encounter or conflict or crisis may destroy or humiliate or compromise them, then they will not take risks or push limits. Perls felt authoritative because he had paid his dues. He took risks. Hence, the confidence of charisma may have a real base.

Part of Perls's trust in himself was his uninhibited streak, the ability to behave like a child one minute and a sage at another, to wholeheartedly flip from situation to situation, stimulus to stimulus, in a way that seemed entirely natural, spontaneous and spirited, and that paid no regard to the opinions or sensitivities of others. As Werner Erhard described him, 'Fritz had class, if one defines class as the ability to graciously not give a damn' (Gaines, 1979, p. 173). There is a lovely story that illustrates this:

> Sometimes when people ask me about the concept of 'living in the now' — what does it mean? — I give them this example from Fritz's behaviour. [One] evening, with

people standing around the table in glittering candlelight, Dale Metzger, who was a veteran of many (of Perls's) workshops and the husband of the woman who had baked the cake, got up, raised his glass, and said, 'To Fritz, for being just the way he is'. Fritz looked around with tears in his eyes; I could tell by his breathing that he was very touched and excited. As we all lifted our glasses, he drank, then said, 'This is the lousiest champagne I've tasted in a long time. What did you pay for it?' Someone told him, and he said, 'Oh my God, we could have gotten so much better stuff for less money'. (Gaines, 1979, pp. 271–272)

The result, as Carter said of Churchill, was someone for whom 'everything under the sun was new', whose 'approach to life was full of ardour and surprise', whose 'mind found its own way everywhere', and who was 'impervious to atmosphere', so radically autonomous was he (Carter, 1965, pp. 18–23). In this regard, Heinz Kohut has spoken of how such self-referential narcissism leads to a severe reduction in the educative power of the environment (Kohut, 1976, pp. 414–415), but it may also be very appealing to others, perhaps reminding them of their own lost narcissism, and of what freedom might feel like without their internalised restraints, introjected values and 'fantasy overlay'. There is something very free and freeing to the charismatic personality; others become more alive around him.

All of this results in the unusual quality of attention that charismatic personalities possess, and which enables them to creatively engage with others in an inspirational way. Whether it is Adolf Hitler listening for the nuances of the jobless German workers, or Germaine Greer resonating with the concerns of women striving for self-fulfillment, the charismatic personality brings to the encounter a creative dimension that is extraordinarily rich.

When Fritz was 'on' he was the most exciting therapist who ever lived. His was the unique ability to pierce down into a person and to grasp what was most basically awry; where grief, fury, death lay deeply hidden (Fagan, cited in Gaines, 1979).

Social Insight

In his interactions, Perls was sometimes described as 'making so much contact it was obscene' (Gaines, 1979, p. x; Shepard, 1976, p. 103).

(The term 'contact' in this usage denotes encountering another person in an especially honest, open-hearted and accepting way; this is discussed further in the concluding chapter.) He was able to keep himself differentiated while using the immediacy of his attention to get right in close with them, and this led to him reading them so well. At his best, Perls had extraordinary powers of observation:

> At [Mendocino] Hospital we called together a group of all the key administrators, and Fritz was going to train them. I remember the opening session very clearly. I was the only one present who knew Fritz at this point. He didn't know any of the others. We were all sitting around, the chief of this and the chief of that, and all these dignitaries; we had worked together for years and knew each other well, and as people we were chatting, Fritz raised his hand and asked for silence. Then he went around the group and told what he could see in each one. He just about blasted us out of our chairs because he saw what we knew to be the real nature of each person. He hit everyone with deadly accuracy. Paul Frey, a friend of mine who has an immense concern with eschatology (the end of all things), is a theologian and a psychiatrist. As an expert in the cataclysmic ending of all things, Frey was concerned with everything, agonising about how the world would be destroyed, and Fritz could see the whole thing there in Paul's face. He said, 'Look at the agony in his face'. When he said it we could all see it. Before he said it, none of us had been aware of it. Then he got to a psychiatrist who started talking nervously, and Fritz just sat back and waited. He gave him maybe ninety seconds and then he just raised his hand and said, 'I feel you are dumping a load of garbage on me', which was actually very accurate. It was verbal garbage. It wasn't worth a damn and there was a great deal of it pouring forth. Everyone was stunned because this particular psychiatrist always did that. It was, in effect, wordy garbage, but no one had ever had the clarity to see it, or the callousness or the daring to say anything like that. I remember another psychiatrist who was a very nervous, ambitious and uptight black. He was just blown out of the room. (Gaines, 1979, p. 62)

When asked to explain his ability Perls simply said, 'I have eyes and I have ears and I am unafraid' (Gaines, 1979, p. 128). Another example from a psychotherapy workshop:

> In half an hour he was doing the work that an analyst would take years to do. He was going very fast and was very efficient ... he had this incredible sensitivity and insight ... Abe and I would sit there ... and Fritz would do something and we'd just look at each other in amazement. Once, for instance, [with] a girl whom Fritz hadn't had anything to do with ... he just took one look at her and he said, 'Did your mother try to abort you?' She was startled. She said, 'Yes, how did you know that?' I remember feeling in that first workshop that I was seeing a series of miracles. (Gaines, 1979, pp. 129–130).

How did he do it? Did he think differently or feel differently? Perhaps. In *Prophetic Charisma* I attempted to describe the extraordinary mentation of the charismatic prophet, a radical freedom of thought resulting from an 'inner spaciousness and wildness of mind' that might account for the uncanny 'presence' that such figures emanate (Oakes, 1997, p. 175). Nowadays, I tend to think of this not so much as a permanent state of mind but as something that is constantly being recreated; it is more a product of the enormous and continuous creativity of such figures, a habitual process that ebbs and flows, that is sometimes 'perfectly on' and at other times not on at all. As another of Gaines's informants reported:

> My experience of Fritz was that he was not doing gestalt therapy; he was creating it at every moment. He had a certain repertoire, but the essential element was to be with what was happening from minute to minute, every minute. (Gaines, 1997, p. 298)

Just as Perls's quality of creative attention gave him extraordinary access to the unconscious states of others, so too Churchill's quality of creative attention gave him access to the unspoken hopes and needs of the British people, as the creative attention of the charismatic CEO enables him to read the ambitions and insecurities of his subordinates. Such extraordinary 'mind-reading' is too well-attested to be dismissed as merely a figment of the followers' imaginations, and it has even been investigated by researchers:

> Years ago a young man was being introduced to his new professional colleagues in the department of psychology; among them was Silvan Tomkins. As the young man elaborated on his many interests, views and intellectual dilemmas he quite exceeded the time that other speakers had taken. Silvan turned to a colleague, lowered his voice and said, 'That young man lost his beloved mother at an early age'. In fact, he had. But when Silvan was queried about his acquaintanceship with the young man he replied, strangely enough, that he had never even met him … Tomkins seemed endowed with a supernatural knowledge of the human mind and its longings. He always seemed to know more about people than was discernible from the observable facts. Indeed, at the memorial service held for Silvan in 1991, not only did renowned psychiatrists single out his uncanny ability to fathom the essential elements of people in a way few could, but even his garbage man described him as a 'yoda' — a wise man. (Magai & Haviland-Jones, 2002, p. 3)

What is most intriguing about Tomkins is not that he had such an extraordinary gift, but that he was unable to satisfactorily explain how he did it. Even though he was a research psychologist and was determined to explicate his 'system', he never really succeeded in making it accessible to others; he could not explain his ability. And despite Perls's writing several books on his therapeutic technique, it is now clear that he too was unable to explain it.

Social insight is the main tool by which the charismatic personality gains power and recognition. He seems to know in advance what his followers want and to lead them to it. He sometimes seems able to even know what they are thinking. His ability to 'read' an audience is so extraordinary as to seem occult, and is treated as such in religious settings, being attributed to a psychic gift. Fortunately, there are enough examples of such phenomenal expertise from more mundane settings to enable us to avoid invoking the metaphysical.

Milton Erickson was another extraordinary psychotherapist who at times seemed almost clairvoyant, and he has left us an account of how he came by such a talent. As a child, he suffered from polio. In order for him to participate in family life he was provided with a large mirror that was placed above his bed, thus enabling him to watch the comings and goings of his family. Through long periods of immobility he studied his family intensely and eventually he became able to predict their behaviours. Later, as a psychiatric intern, he studied hypnosis and he used the knowledge he had gained as a sick child to develop the approach to hypnosis that bears his name. He was able to read the inner states of others with phenomenal accuracy, and to subtly guide them towards healing (Bandler & Grinder, 1975).

Creativity

As Aristotle observed, creativity seldom exists without a degree of madness, and creative types are invariably inclined towards melancholy (Ochse, 1991). The connection between creativity and psychopathology is now fairly well-established, but quite which psychopathology, and quite how and why, remains mysterious (Andreasen, 1987; Becker, 2000–2001; Hendricks, 2004; Jamison, 1993; Ludwig, 1995; Post, 1994). Bipolar disorder is the most fashionable candidate currently (Gartner, 2005; Hershman & Lieb, 1988, 1994, 1998; Jamison, 1993), but a recent review included narcissism and the closely related disorders of megalomania, hypomania, paranoia, borderline personality and psychopathy as being also associated with creativity (Brod, 1997). There is much overlap between these disorders; for example, between borderline personality and bipolar disorders (Gunderson et al., 2006). However, it is also entirely possible that the main connection may simply be that these disturbances provide tremendous arousal and stimulation; the artist who suffers from bipolar disorder or from a personality disorder may have much greater energy than a 'normal' artist.

An important Australian study has explored these relationships (Hendricks, 2004, p. ii). It compared artists with 'normal' controls, and found that the main factors linked to creativity were 'healthy rebellion' and a 'predisposition towards psychosis'. Artists scored higher on a range of pathologies, including several personality disorders, and for all of the subjects in this study, narcissism was related to creativity. Intriguingly, it was the healthier aspects of narcissism —

authority, superiority, self-sufficiency and vanity — that were most closely related.

This suggests an answer to the question of why some disturbed persons manage to tame their disturbance but others do not. It has long been assumed that adjustment is a matter of the degree of severity of the disturbance, all else being equal. By this logic, the more disturbed the person is, the less chance he or she has of adapting to their pathology. But perhaps it is really the creative ones who adapt, irrespective of the severity of their disturbance, whereas the less creative ones find it harder to cope. This would confirm the thesis of Victor Bloom who, in an article on psychoanalysis and creativity, has described how personality-disordered patients may 'save' themselves through their creativity (Bloom, 1998). Hence even severely disturbed individuals may redeem themselves with intense creative efforts.

However, it is not just intellectual or artistic creativity that charismatic personalities possess, the kind of applied problem-solving aimed towards a particular goal. It seems to lie deeper in their beings. The sense of novelty and freshness that they bring to their perceptions, which Carter so eloquently described in Churchill, has no aims, it just is. This suggests the presence of something more primitive and rapturous, as was portrayed by author Nikos Kazantzakis (1971) in his novel *Zorba the Greek*:

> While going down a slope, Zorba kicked against a stone, which went rolling downhill. He stopped for a moment in amazement, as if he were seeing this astounding spectacle for the first time in his life. He looked round at me and in his look I discerned faint consternation. 'Boss, did you see that?' he said at last. 'On slopes, stones come to life again.' I said nothing but felt a deep joy. This, I thought, is how great visionaries and poets see everything — as if for the first time. Each morning they see a new world before their eyes; they do not really see it, they create it. (p. 140)

Rollo May has described the subtle and paradoxical attitudes that underlie such creativity (May, 1975). First, there is detached involvement, a cool head directing passion; emotions at a distance. Then there is mindless perception, the free play of perception without the imposition of categorical thinking; the suspension of judgment.

Delayed closure, an attitude of searching and analysing for its own sake without any rush to problem-solution is also a factor, as is converging divergence, the interplay of lateral thinking (De Bono, 1970) and linear logical processing. Other components include relaxed attentiveness, flexible persistence, confident humility and a degree of selfishness. Finally, and perhaps most important, is constructive discontent, an emotion-based need to oppose and improve. Charismatic personalities are much more than mere fault-finders; rather, they are famous for their 'divine discontent'. The moment they reach some resolution they attempt to find ways of going beyond it, and often, even after arriving at their 'truth', they still try to look beneath or beyond it.

The creativity of charismatic personalities has been studied directly (Hall, 1983; Labak, 1972), and also considered by leading theorists (Conger & Kanungo, 1988). It has shown up in several areas of functioning including relationships, work and managing people, but the charismatic personality appears to be most creative when devising a 'strategic vision' to inspire and lead others, that is, when oriented towards narcissistic ends.

Heinz Kohut considered creativity in his essay on Sigmund Freud, and again in his posthumous piece, 'On Courage' (Kohut, 1976, 1985). He described how intense creativity may deplete a person, leading them to lean on a charismatic personality for support. For charismatic personalities, their creative agenda becomes a 'strategic vision' that consumes their energies. In the words of Wilhelm Reich it is a 'life-affirming flame' (Sharaf, 1983), that may even be perceived as another identity, a muse that drives one mad if not obeyed. This involves the translation of childish narcissism into a utopian project that requires the recruitment of others for its fulfillment.

Memory

Memory is central to all cognitive functioning so it is no accident that it is highly developed in charismatic personalities. The phenomenal memories of charismatic figures have been discussed by several theorists, and the topic routinely crops up in biographies (Willner,

1984, pp. 144–146). By developing their memories charismatic personalities gain great leverage over themselves and others. There may be a genetic component to this, but there need not be. In an extraordinary experiment, researchers were able to train a student of average intelligence and memory to improve his 'memory digit span', that is, the number of digits he could recall after seeing them briefly, from 7 to 79 digits. This took approximately 12 months to achieve, and there was no suggestion that this was any kind of upper limit. The student's ability to remember the digits after the experimental sessions also improved enormously (Ericsson, Chase, & Faloon, 1980).

The developing narcissist trains his memory to extraordinary levels so as to out-perform others, and to appear to be an exceptional person. A great memory enables him to be many different things to many different people; to remember names and conversations long past, and to recall intimate details of past encounters that others have forgotten. This enables him to court and charm others, to appear to read their hearts and minds, and to seem to be more empathically involved with them than he really is. His memory is in the service of his ambition; it helps keep him one step ahead of the crowd.

Phenomenal memory capacity is not all that unusual, and it has been studied. So-called 'photographic memory', while rare in adults, occurs in about 8% of children. It seems to peak shortly before puberty and to decline sharply after that (Haber & Haber, 1964). Quite why it is retained by some and not others remains a mystery, although obviously it may become reinforced by the rewards it bestows, as occurred with Churchill. However, when considering charismatic personalities, another possibility suggests itself.

In an essay discussing adaptation, psychoanalyst Helm Stierlin has noted the capacity of infants to read the emotional states of adults. They have to; they have little else to go on prior to gaining language. He describes the infant as a super-specialist in understanding unconscious states, able to perceive in a particularly clear and immediate way the feelings and moods of others. He relates this ability to primary narcissism, saying that, 'the undifferentiated child also has capacities for obtaining and organizing data that most adults have lost' (Stierlin, 1959, p. 148). This linkage between the undifferentiated state of primary narcissism and extraordinary cognitive performances implies that they may go together. Thus, when asking where charismatic personalities get their prodigious talents from, especially their social insight, creativity, mental detachment and their phenomenal memories, we should consider the possibility that by retaining their infantile narcissism, if only partially, they have also retained archaic modes of being that later develop into these faculties. It is a complete package.

Recent research on savants illuminates this. Savants are individuals with significant social or cognitive disabilities, usually autism or Asperger's syndrome, yet who possess phenomenal abilities in some other domain such as memory or mathematics or music. While not all impaired children become savants, as many as 30% do develop some kind of special ability (Biever, 2009). It now appears that several traits underlie the development of savant skills. Attention to detail leads savants to focus more intensely on the ability that catches their interest, for example, memorising and obsessiveness enable them to analyse it endlessly in order to draw out every possibility within it. Tireless practice refines and develops the ability to extraordinary levels, and motivation or will is a big factor. As one researcher has described it, 'The survival instinct gets turned with extraordinary force into something else ... When people see [it] they think it is amazing, almost religious. But to me it's mainly just hard work' (Biever, 2009, p. 41).

This probably explains the extraordinary abilities of the charismatic personality. Whether it be memory or manipulativeness, social insight or creativity, the bottom line of charismatic prodigality is likely to be 'just hard work'.

Charismatic Communication

Charismatic communication is a form of speech that addresses issues beyond or beneath whatever is ostensibly the focus. It involves an orientation either to ultimate concerns, or to a more in-depth dimension, sometimes to both. It relates whatever is the focus to greater or deeper matters. If the speaker is a virtuous leader, his charismatic communication inspires hope and evokes higher virtues in the listener. President John F. Kennedy's statement 'Ask not what your country can do for you, but what you can do for your country' is a typical example. But it can also be used to generate hate, as when Adolf Hitler aroused the crowds by his depiction of the noble German nation 'on its knees' before international Jewry. He may have begun by talking about unemployment and inflation, but he soon moved on to the 'volk' and its 'destiny'.

Charismatic communication is not used solely in public speech-making to evoke hope; it also occurs more intimately, one-on-one. At this level it is more likely to play upon universal fears, touching emotional bases that underlie whatever concerns occur on the surface. Consider the following scenario.

A 40-year-old man visits a therapist. The man's eyes look tired and he walks slowly. After sitting he pauses for a while, sighs sadly, then softly says, 'My wife has left me'. A common therapeutic response to such a presentation is paraphrase and elicitation, typically: 'Your wife has gone. Hmmm. Can you tell me more about that?' Another response might be to reflect, and then to meta-comment: 'You say your wife has walked out. Hmmm. You sound pretty sad'. Or the therapist might reflect and add an element: 'Hmm, your wife has left you. Was it a great shock?' Or the therapist might reframe: 'So your wife left you. Yes, that can happen. Tell me more'.

All these responses involve common therapeutic fare in that they meet the client's concerns at the level at which they are expressed, basically stimulating further exploration of the client's issues while steering things this way or that. They express solidarity by feeding back what the client has said, and they show attentiveness by accurate paraphrasing. They may provide additional material such as, 'You sound pretty sad', or they may seek specific elaboration by asking, 'Was it a great shock?' Or they might normalise by reminding that this kind of thing happens to many people.

A charismatic response would attempt to orient what the client has said to some underlying and so far unstated universal fear, typically, fear of rejection, abandonment, being unlovable or the like. For example, the therapist might pause, then observe, 'So your wife has left you. Hmmm. And now you fear the worst?'

This response has wide implications. First, it exposes and acknowledges what is probably the client's deepest fear and the main source of his pain, the fear that without his beloved wife, his life has no meaning or hope and that he may never be happy again. This acknowledgment provides extraordinarily close empathy. It feels like the therapist has read the inner workings of the man's mind, seen into his deepest fear, perhaps even before he himself has become aware of it. This builds a sense of union. The man feels seen and understood in a very intimate way, precisely the way he most needs to feel at this moment. (Actually, he might not be ready yet to address such depth; this is an illustration only.)

Second, it implies hope. It suggests that there may be other possible ways of handling this event. A truly hopeless remark that also drew upon a universal fear would be, 'Well, your life is over now'. But the words, 'you fear', qualifies the comment, turning it into a question. It implies that there may be other ways to see things, for what one fears is true may not actually be true. This allows for hope, tenuous and hesitant, but hope nonetheless.

Third, it implies universality. To make such a comment the therapist would have to be very familiar with this phenomenon, presumably because he has seen it often. He seems to know that after the loss of one's beloved, of course life appears hopeless, and of course one descends into primal fears. This understanding connects the sufferer to the community of sufferers everywhere, down through history, who have suffered similarly and it removes the stigma of aloneness. The client's pain is universal; it has not happened just to him, or because of some failing unique to him.

Fourth, it implies survival. By connecting the client to the world-wide community of sufferers, the comment implies that, just as others have survived such pain, so too will the client.

Fifth, it implies an unexpected richness to life despite this loss. By implying that the client may be wrong in assuming that without their beloved life is meaningless, it suggests that perhaps he is wrong about other things as well. Life may be far richer, complex and anomalous than he is aware. It locates the man's suffering in one arena of life, albeit a major one, but it reminds him that there are others. Again, the client may be hurting too much to want to hear this yet, but the suggestion will linger, and when he is ready to hear it he may do so.

Sixth, it is unexpected. The comment has an element of shock value, and there is a sense of feeling the universe opening up and revealing itself. The client begins by talking about something that usually would engage a listener, but the therapist seems to stand at one remove, then to change the focus according to some broader or deeper view, and to point to other things while acknowledging the client's issue. This is surprising and potentially liberating. There is a sense of the scales falling from one's eyes, as one follows the therapist's lead into existential and relativistic areas by 'thinking outside the square'. The client may even wonder what else the therapist knows about him. Charismatic communication makes several assumptions. It assumes a degree of universality to human experience. As Kluckhohn and Murray's classic dictum had it: 'Every man is in certain respects (a) like all other men, (b) like some other men, (c) like no other man' (Kluckhohn & Murray, 1953, p. 53). By reminding listeners of this universality, and therefore referring them back to their community or nation, the speaker bonds them to himself and locates both him and them within something greater. At such times the leader speaks the language of unity, and of their shared greater identity, rather than of their individual egos. There is a feeling of womb-like inclusion.

Charismatic communication also assumes the existence of ultimate concerns and noble aspirations. As the historian of religion Wilfred Cantwell Smith has shown, everywhere throughout history humankind has been aware that we live in a world whose greatness transcends our grasp but does not totally elude us; that truth, beauty, justice and love 'beckon us imperiously yet graciously' (Smith, 1979, p. 130). By reminding the listener of these eternal verities, the speaker lifts him out of his individual ego and portrays him as an historical actor within the great inclusive sweep of history and progress.

Finally, charismatic communication depends on understanding and evoking universal fears to do with meaninglessness, loneliness, isolation, rejection, abandonment, death, chaos, powerlessness, helplessness, annihilation and engulfment. This gives it a 'straight for the jugular' flavour that is paradoxically both unnerving and releasing at the same time. It is unnerving because it signals that the speaker can almost read one's mind and knows one's deepest fears; the anxious question arises: 'What else might s/he know about me?' But it can be releasing if it shows that the other is comfortable with such fears; the worst becomes survivable.

Charismatic comments also depend on one's theory of life and human nature. For example, Christian evangelists might invoke fear of separation from God (Hell), or rejection of God (sin) as one's deepest fear, but to a nonbeliever such matters may be meaningless. This is why one person's charismatic is another's bore. To be effective, charismatic communication must reflect the listener's worldview and values.

There is a quality of fearlessness to charismatic statements. The therapist making a charismatic comment shows that they are prepared to flout social niceties, to 'cut to the chase', and this signals a substantiveness on their part rather than mere conformity. There is a freedom here too, like the child speaking a truth that adults are too inhibited to acknowledge, the wisdom of innocence, which suggests a more holistic and natural way of being. These, in turn, open up a heroic dimension in which one rises above convention and adopts a richer perspective on one's problems.

Charismatic communication is only possible where there is accurate perception. One needs to see inside the other's world-view and to stand in their shoes, to identify the concerns that drive them. In the above example, if the client had appeared before the therapist in a happy state while announcing that their wife had left, then the therapist would have responded differently, perhaps merely by asking how the client feels about what has happened. Or, if the client appeared angry the therapist would also respond differently, perhaps meta-commenting on the anger. If the therapist did not read these states accurately his comment would either be dismissed as silly by the client, or it might unleash the anger.

Hence the speaker must accurately perceive what their listener is experiencing, as well as some aspects of their context and being, in order to construct a comment that hits the mark. In the above example, the therapist might detect from a certain stiffness and controlled exterior that the client is one of those men who are unable to admit to weakness, in which case the therapist would remove any reference to fear from their reply, inquiring perhaps, 'And now it seems like your life is over?' Or the therapist may suspect from the client's fragmented behaviour, dilated pupils and disheveled appearance that his ego strength is unable to endure such frank talk, at least for now, and thus he might choose to avoid all comments that would increase this distress, perhaps softening the message with a euphemism and saying, 'You seem near the end of your tether'. Or the therapist may suspect that something does not quite add up, that the client's distress is more appropriate to the loss of his cook, cleaner and bedroom slave than the loss of a life partner and soul mate, in which case the therapist might respond differently again, perhaps by asking, 'Where does that leave you?' These responses might, in the right circumstances, be accurate and empathic, compassionate and moving towards depth, and they share some nuances, but they lack the 'full-on' quality of the charismatic comment that goes unequivocally to one's deepest concerns.

To create and deliver a charismatic comment is a very high-level skill, for it involves saying what the client may be too defensive to hear in a way that gets under their defences. A good example comes from a training workshop given by Bert Hellinger, a well-regarded Dutch therapist. The participants had been invited to present cases for discussion. One worker said that she was at a loss to know what to do about a particular patient at a hospital she worked in, who seemed to get on well with other staffers and patients but not with her. Hellinger paused for a moment, then asked, 'Are you competent?'

Hellinger (1998) was not putting the trainee down, nor triumphally revealing her insecurities, nor merely provoking her. His words were delivered with concentration, yet matter-of-factly, and as if the answer was all-important. The question went unerringly to her deepest fear in that context; that there might be some failing within her. He was asking a question that connected with her doubts about herself, tuning in to her process. By respectfully giving her his full-on attention he showed that she really mattered to him, and this implied that she was safe to explore with him her doubts about herself. He was confronting her with her deeper fears in a way that got beneath her defences, joining with her in a journey towards her unknown regions.

Charismatic personalities seem able to access such strategies even in the heat of the moment. Bhagwan Shree Rajneesh, the deceased Indian guru, was at his ashram in Poona one day giving darshan (blessing) to a large group of his followers when suddenly a stranger leapt to his feet brandishing a gun. This man pointed it at Rajneesh and shouted that he was going to kill him. Rajneesh looked the man in the eyes and asked, 'And after you shoot me, will you be happy?' The man broke down in tears and was led away. Such presence of mind in charismatic figures, the ability to look beneath or beyond the immediate situation and to return with a hopeful or disarming comment even in their extremity, has also been attributed to Mohandas Gandhi, Franklin Delano Roosevelt, Kemal Ataturk, Benito Mussolini, Kwame Nkrumah and Charles de Gaule (Bass & Riggio, 2006, p. 67; Wilner, 1968, 1984).

An example in which both conversationalists use charismatic communication comes from the film *A Beautiful Mind*. It has the schizophrenic hero of the film, Nobel prize-winning mathematician John Nash, in conversation with an intelligence operative named Parcher, who begins by complimenting Nash on his earlier mathematical work.

> P. Impressive work at the Pentagon.
>
> N. Yes, it was.
>
> P. Oppenheimer used to say, 'Genius sees the answer before the question'.
>
> N. You knew Oppenheimer?
>
> P. His project was under my supervision.
>
> N. Which project?
>
> P. (Silence.)
>
> N. That project.
>
> P. It's not that simple you know.
>
> N. Well, you ended the war.
>
> P. We incinerated 150,000 people in a heartbeat.
>
> N. Oh, great deeds come at a great cost Mr Parcher.

P. And conviction, it turns out, is a luxury of those on the sidelines Mr Nash.

There are so many reorientations of direction and depth and focus in this conversation as to defeat a simple analysis. There are at least eight shifts of direction in the piece as it leapfrogs over immediate concerns to plunge ever deeper into fundamental issues, or broader to include widely different or transcendent domains into the narrative. In a mere 75 words it spans the ethics of war, the nature of ultimate weaponry and the relativism of individual conscience, while informing each participant of the other's sympathies and gaining a measure of both.

Will

After all the theories of charisma have been considered, it remains hard to avoid the suspicion that beneath all the layers of description and explanation there exists something more fundamental. The intensity and drive of charisma seem to come from somewhere else, some unknown region of the mind that powers the charismatic personality forward to survive difficulties that would defeat others, to drive him or her with a seemingly irresistible flow of energy. This feeling is strongest when one is actually watching such a leader close up. There is a sense of continuous activity, as if some larger-than-life process is at work, and of some inexhaustible source of energy and inspiration.

This power is will, a force that has been widely discussed in philosophy from classical times, but which has been strangely overlooked by psychologists. 'Will' has been defined as the prime mover of the organism, something more basic than intention, hope, desire or any of the other overlaid motivations, all of which depend on the will for their energy.

Leslie Farber (1961, 2000) was the leading psychological theorist of the will. He described it as a 'residue' that remained after psychological reduction. For example, one might explain Leonardo Da Vinci's painting of the Mona Lisa in terms of his genius, creativity, natural talent, social aspirations, need for appreciation and money and so forth. One can also cite his unconscious desire to impress his parents, or his fear of failure, or his quest for immortality. But after all these explanations have been factored out, will remains as an underlying force, the power that surged Leonardo forward in his life. For various situational reasons he became an artist at a particular time and place. But the nature of his will meant that his strivings and creativity were of a higher order than his contemporaries, and this ultimately led to him producing the Mona Lisa. Thus, will is a kind of engine in the psyche.

Farber notes that will may be conscious or unconscious. It may be a spontaneous, Zen-like 'flow' of optimal experience in, say, a superlative tennis game (Csikszentmihalyi, 1991), or it may be a deliberate 'act of will' in which one focuses on performing some methodical action, perhaps when one's tennis game goes wrong. In a psychiatric sense, will is the opposite of basic anxiety, that 'feeling of being small, insignificant, helpless, deserted, endangered, in a world that is out to abuse, cheat, attack, humiliate, betray ...' in Karen Horney's memorable account (Horney, 1937, p. 92). Farber treats hysteria as a disorder of the will; its principal expression is wilfulness and obstinacy, and it is perverse and without reason, full of self-assertion yet empty of self.

Will seems to be that part of the organism that expresses the life force, and that drives one to engage with life and to achieve one's potential. Thus it is not emotionally neutral, for the life force is imbued with love; as Thomas Dow argued, and as Benjamin Zablocki and Raymond Bradley showed, charismatic movements are largely experiments in love. As an expression of the life force, charisma may be viewed as originating in an abundance of will.

Will obviously varies from person to person, and it is probably also an inherited trait.[1] This might explain why different individuals in comparable circumstances develop in different ways; why some prosper and others languish. Most children raised as Churchill was go on to become ineffective adults lacking in confidence, initiative and creativity. This was even recognised among the aristocracy of his time, and was widely thought to be the usual fate of second and subsequent sons (his father conformed to this pattern). However, Churchill's powerful will enabled him to avoid being depleted and demoralised by his sufferings, and to devise saving explanations for his troubles; explanations that soothed and empowered him.

There is an excellent account of how this may happen in the life of psychotherapy pioneer Albert Ellis (Ellis, 1972, 1996; Magai & Haviland-Jones, 2002; Weiner, 1998). Ellis suffered great neglect as a child. He was sickly and he underwent several traumatic early hospitalisations. These experiences stimulated him to develop the fundamentals of what he later called Rational Emotive Therapy by devising the strategies that enabled him to cope with adversity and to endure painful emotions. He devised, and then drilled into his head, basic formulations such as, 'Life is full of hassles you can't control', and 'What does happen could always be worse', and 'Wait before you panic'. We may recognise in these 'detoxifying scripts' (Tomkins, 1987) a cool realism and a precocious absence of sentiment that less wilful people may not possess.

If Ellis is to be believed (and none of his commentators suggest he is not, although it seems extraordinary), he was constructing his tough-minded philosophy of life as early as age four. This may also explain why and how, in a neglectful family with few reading materials, he taught himself to read before he went to school. Certainly intelligence of a high order must also have been involved, and no doubt several other positive talents. But his intelligence and talent could easily have been turned inward and absorbed into self-pity. Instead, Ellis defended himself from his vulnerable feelings with his scripts. Later his life involved constant activity; even as a 75-year-old he routinely worked up to 80 hours each week. He became emotionally isolated, but was also driven, ambitious and fascinated by psychological issues. In time he founded a psychotherapy movement, attracted a large and loyal following, and helped many. He lived his life through his audience, giving advice on all aspects of work, marriage and parenting. In his long career he received numerous scientific and professional awards. He died in 2007 at the age of 93.

Brain Research

The development of sophisticated techniques of brain research, especially Magnetic Resonance Spectroscopy, have opened up new possibilities to investigate the biology of traits and behaviours. Promising lines of research relevant to an understanding of charisma include the discovery of 'mirror' neurons, and the condition known as 'low brain masculinisation'.

So-called mirror neurons are brain cells that have been implicated in self-consciousness, intersubjectivity and empathy (Gallese, 2001). They activate when one observes another person doing something that one has also done. Given the extraordinary social insight of charismatic figures, and the contradictory nature of their empathy, which Heinz Kohut describes as highly accurate when it serves their interest but stunted otherwise, it appears likely that investigation of the functions of mirror neurons and their relationship to social behaviour may shed much light on the actual mechanics of narcissism, and by extension, charisma.

Low brain masculinisation occurs when, during development, the male brain is exposed to lower than usual levels of the male sex hormone testosterone. The effect of testosterone on the developing brain is to enhance masculine traits; individuals who are exposed to high levels of testosterone tend to be very aggressive with relatively poor social awareness. Low brain masculinisation leads to reduced aggression and to the development of more typically feminine traits such as social sensitivity. It appears likely that low brain masculinisation is what has enabled civilisation to arise; our primate ancestors were both more highly masculinised and more violent than we currently are, and they lived in rigidly hierarchical social systems dominated by alpha males who jealously guarded their harems.

Low brain masculinisation also appears likely to be related to several traits closely related to charisma, especially creativity and social sensitivity (it has also been implicated in homosexuality; Berman, 2003). Theorists have suggested that charismatic leadership demands a combination of both masculine and feminine sensibilities (Hackman, Furniss, Hills, & Paterson, 1992), and in the case studies presented in this book the theme of softness recurs. Of course, all such influences are mediated by interaction with the environment, so there is nothing binding about low brain masculinisation and its influence on charisma. Further, such brain studies as might clarify matters are still in their infancy. But it appears likely that low brain masculinisation may be involved as a necessary, though not sufficient, contributor to the development of charisma.

Concluding Remarks

None of the above should be taken to imply that a particular set of traits determines the development of charisma. This chapter has canvassed several traits that seem to be promising for further investigation because they appear to be central to charisma. However, the point should not be overlooked that they do not constitute a 'talent' for charisma. Many people have these abilities but do not develop charisma. In the authoritative *Cambridge Handbook of Expertise and Expert Performance*, Anders Ericcson has advised:

> The search for stable, heritable characteristics that could predict or at least account for superior performance of eminent individuals [in sports, chess, music, medicine, etc.], has been surprisingly unsuccessful ... Systematic laboratory research ... provides no evidence for giftedness or talent. (Ericcson, 2006, p. 42)

Expertise in any arena comes from persistent, applied practice, and especially from focused practice of a trial-and-error nature in which abundant feedback about effectiveness is integrated into one's efforts on a regular basis (Dawes, 1994, p. 111). Charismatic personalities mostly develop their prodigious abilities through sheer hard work. Any attempt to explain charisma through recourse solely to a set of exceptional innate talents is misguided.

Endnote

1   See *New Scientist*, (2008, November 1), 'In brief: If you've got that precious mettle, thank your parents', p. 16.

● ● ● ● ●

PART **THREE**

• • • •

Conclusion

CHAPTER FIFTEEN

'He was a monster, but …'

We may, it seems, be fated to be governed by leaders with wills to power based on unconscious fantasies of omnipotence; leaders whose conviction can be harnessed to constructive ends, but whose effectiveness is always impeded by the delusion that they, and they alone, can, and must, control the work of government.

James Walter

Charismatic personalities may be the loneliest people in the world. Even the successful and self-fulfilled Nelson Mandela, according to his former secretary Mary Mxadana, felt depleted and exhausted when deprived of narcissistic supply. Fritz Perls also seemed to carry a burden of loneliness.

> On the last night of the conference we had a party in our room, no different from every other convention party, too many people crowded into a small room, babbling about too many things. [Then] Fritz was suddenly there and suddenly something was different. The noise level dropped; people were trying to be serious. And goddamit, I found myself idolising him just like everyone else … A spontaneous ritual began to ensue, of individuals cautiously approaching Fritz, speaking a few questions, and respectfully backing away … Fritz walked out. Feeling responsible I chased after him and apologetically stammered something like 'I guess gods aren't supposed to be at cocktail parties'. In that deep, scratchy, warm voice he replied, 'I am Fritz, I am not a god'. For a moment we spoke. I mentioned his fame, the adulation he was receiving, the fact

that he was finally being heard ... He said that it was not really he that was being heard, not really Fritz that was getting famous, but only 'who they think Fritz is'. I was staggered with his loneliness. I was, and remain, aware of the paradox of a man seeking his entire life to be heard, to be listened to, and then when it seemed to be happening, was left feeling isolated and alone by people only hearing and respecting their fantasies of who they thought he might be. I was overwhelmed with a sense of recognition of him. There was nothing to say. We hugged. I never saw him again. (Gaines, 1979, pp. 391–392)

Another charismatic personality, Joseph Smith, founder of the Mormons, towards the end of his life, addressed this problem directly:

In a sermon several months before his death, Smith expressed profound frustration at his inability to be understood, to get his deepest message across to even his closest followers. He declared, 'You never knew my heart. No man knows my history. I cannot tell it. I shall never undertake it. If I had not experienced what I have I should not have known it myself.' (Foster, 1983, p. 98)

Both Perls and Smith claimed to have been struggling to have their messages understood all their lives. But they only formulated their messages in mid-life. Their earlier struggles must have arisen because their narcissistic personalities limited their abilities to connect. They may even have longed for understanding since childhood. As young men they would have found themselves out of sync with their fellows, and unable to be accepted as 'one of the boys'. But they refused to be defeated by such difficulties. Each worked hard to learn how to charm and manipulate their way towards acceptance. Stimulated by their painful distance from others, yet finding it bestowed certain advantages on them, they became exquisite students of human behaviour. They learned to inspire others, and in time they emerged as leaders. Each fashioned a great world-saving ideology which, when accepted by their followers, elevated them to the status each felt he deserved.

They were not insincere; they constructed their messages in the hope that by providing solutions to human suffering they would themselves attain redemption. They tailored their messages to the sensibilities of their followers, refining their teachings many times. They enjoyed great success; Smith communed with angels and created the largest modern religion, and Perls founded a major psychotherapy school, and even claimed to have attained perfection (Stevens, 1970, p. 54). They received much applause and idealisation from their grateful followers. But they never felt fulfilled because their own sufferings went much deeper, into their earliest lives, and were not amenable to narcissistic supply. They found themselves trapped. Despite their successes, each lived a life of quiet frustration, and approached his end feeling misunderstood. Each had a tremendous ability to 'contact', but little ability to 'connect'.

Contact is the ability to deeply attend to, and empathise with, another person. It is a profound interpersonal skill which, when employed sensitively, enables one to speak the language of another's heart. It has been beautifully defined by Virginia Satir:

> The greatest gift I can conceive of having from anyone is to be seen by them, heard by them, to be understood and touched by them. The greatest gift I can give is to see, hear, understand and to touch another person. When this is done I feel contact has been made. (Satir, 1976)

In contrast, connection, in the sense intended by E.M. Forster (1995) in his classic novel *Howard's End*, involves three main elements: (1) internalisation of the other within oneself, (2) investment of oneself in the other and (3) commitment over time. The hero of Forster's novel, Henry Wilcox, is a rich capitalist who re-evaluates his life after a tragedy. After much struggle he learns to 'only connect' with those beneath him whom he had previously been contemptuous of, eventually coming to see them as like himself and equally deserving.

The extraordinary ability of the charismatic personality to contact and inspire others co-exists with an inability to genuinely connect with them; to recognise himself in them, and they in him. This is the psychological state of being lonely in a crowd, and for some it is a condition they never escape. In every group he finds himself in, he feels compelled to re-enact the old scenario derived from his primal group and repeated compulsively ever since, in which he strives to dominate. He watches others closely, attending to their attention, observing how they join and depart, waiting his opportunity. He identifies the highs of each member's hopes, and the lows of each one's fears. He especially notes their needs to belong, and how each reacts to threats to their belonging. He sees how, even prior to meeting, they form above-or-below relations with each other. He notices how periodic waves of enthusiasm and anxiety sweep through the group. He thinks deeply about these matters.

He realises that without leadership, most will dither and bicker and achieve little. He sees that despite their denials, most yearn for a leader to lift the burden of freedom from them. They take comfort from feeling that someone strong and wise is in control; he notes the rebel's abiding sense of inferiority and resentment. He learns that power is paramount, that principle counts for little and that strength of personality will usually win out in the end. Ultimately, even reason must be supported by force. He begins to steer the latent irrationality of the group by exploiting its collective hopes and fears, and by recruiting one or another to perform various roles for him. Occasionally, he reflects that if these people could only cooperate more readily there would be no place for him.

In time he finds that his world-view has changed. He comes to think of society's institutions as really just cultic groups where strong personalities dominate. He sees the professions as merely vehicles for the ambitious. He observes the strong farming the weak and being, for the most part, lousy farmers. He sees social progress as enacted and mediated by battling elites. He agrees with anthropologist Geza Roheim, as cited in Becker (1971, p. 146), that culture is merely, 'the fabrication of a child afraid to be alone in the dark', something to be cut through and discarded when he embarks upon his great calling. Yet through all his observations and ruminations there runs a small lonely voice crying, 'I am nothing, and I should be everything!'

This book was prepared against a background of increasing awareness of climate change, the energy crisis and the economic collapse of 2008. Such huge challenges will demand very sophisticated responses, and although the technical solutions are well known, at base these are human problems that will require human solutions. This means leadership. In times of crisis people are more likely to turn to charismatic leaders, and selecting the right leader becomes crucial. Yet, surprisingly, there has been strong debate over whether charismatic leaders really are more or less effective than comparable non-charismatic leaders (all else being equal). In his critique of Max Weber's theory, Raymond Bradley (1987, pp. 29–48) argued that most of Weber's propositions regarding charisma could be shown to be flawed and that, contrary to Weber's assertion, members of charismatic groups took more time performing communal chores and tasks than did those in non-charismatic groups; that is, they worked less efficiently. But like so much about charisma, the resolution of this issue may lie in context. A charismatic leader almost certainly is more effective than a non-charismatic leader at getting certain group tasks done, but he operates in a context of crisis where the risk of failure is great and where he is likely to be closely monitored. A non-charismatic leader is probably less effective at getting those same tasks done, but because he operates in a non-critical context there is less risk of failure and he will probably not be scrutinised so closely. The corollary is that once the crisis has passed it is best to relieve the charismatic leader of his post, or he might maintain the group in a perpetual state of crisis. That said, what advice can be given for assessing charismatic candidates for leadership?

Most of the advice that can be given is the same as that concerning non-charismatic candidates, but perhaps the first caution is to address the hopes and motives of the electorate. Charismatic personalities are usually only selected to deal with severe problems, but by the time a charismatic leader is appointed the problem may have already become too extreme to manage. This is the most frequent reason for charismatic failure, and it has been a recurring theme in many Third World, postcolonial societies. But such severe problems are caused by the community, and for all his inspiring rhetoric the leader can only achieve what is humanly possible. We should not have higher expectations of him than we would of ourselves, and especially, we should not expect him to easily fix problems that we have irresponsibly allowed to escalate.

Next, scrutiny should also be extended to the institutional restraints in place; do not just evaluate the leader, examine the entire context. Post-World War I Germany was a weak target and easily subverted, but most — though not all — modern democracies are stronger. New Zealand Prime Minister, Robert Muldoon, did great damage to his country's economy and polity, and while this could have been avoided by a more astute electorate, nevertheless the institutions he faced, the system of checks and balances, the separation of powers and a muscular media, were able to eventually contain him.

Consideration should then be given to the strength of the candidate's relational restraints: his marriage, family, and family of origin. These supports should be ultimately what he is working for, and a candidate who has only tenuous intimate supports may feel little or no restraint in a crisis. These relationships help keep him on track when the track has been washed away. Two of the most infamous cult leaders of recent times, Chuck Dederich of Synanon, and 'Moses' Berg of the Children of God, behaved quite responsibly until the deaths of their wives, after which they spiralled down into nihilism.

Concerning his psychology, it should be appreciated that the mere existence of pathology is not by itself a good guide, unless it is severe. Many great leaders have had mental health problems (Chafe, 2005). Of course, a very disturbed personality is unfit for leadership, but a candidate who has had a problem with depression or alcohol may still be the best available, as may one with narcissistic tendencies. Contrary to popular belief, not many charismatic leaders self-destruct, and they may sometimes make better leaders than more healthy rational or traditional authorities. But they are change agents, and change never comes without a struggle, so the stakes and pressures are higher and may take a toll.

It may be more helpful to inquire whether his childhood was basically benign. It may have been rough, as Bill Clinton's and Winston Churchill's childhoods were in different ways, but the question is: did he emerge from his family of origin with prosocial values, or did he emerge brimming with hatreds? The angry crusader was angry before he became a crusader, and may stay angry.

A final important question is: do people get hurt around him? Bill Clinton never broke the law, but he accumulated a history of narrow escapes, with marital infidelity a recurring theme, and numerous disillusioned campaign workers in his wake. Then again, Clinton had a stable family and a history of success. Clinton will probably be judged by history as a potentially great president whose personal failings limited his effectiveness, but he did stall the advance of the ultra-right, and his failings were visible long before he became president. Charismatic personalities rarely change their tunes; Adolf Hitler was writing about a 'final solution' long before he achieved power, and much of his public speaking was spent attacking people beneath him. He aimed his rhetoric at the weak rather than the strong, and proclaimed the virtue of hate. Leaders who advance themselves by demonising others, especially those beneath them, are unlikely to stop doing this after they gain power.

Finally, how should we evaluate such individuals? Obviously, according to eternal verities, great achievement is no justification for unethical behaviour. But charismatic personalities often operate outside normal conditions, where the usual rules do not apply. They may find themselves having to make terrible decisions. Their courage is that they are prepared to place themselves in such situations, but they may end up with blood on their hands. Equally, they may be so constrained by their situations that they are obliged to behave in an ethical manner when otherwise they might prefer not to. They may come to be credited with vices or virtues they lack.

Again, much depends on context. This is reflected in the ways different communities evaluate charisma. Charismatic scientists such as Nobel Prize winner Carlo Rubbia, while perhaps not popular, are nevertheless greatly admired in the scientific community for their ability to force their discoveries onto an unsympathetic and resistant world. But for all practical purposes a scientific discovery is either true or false; the charismatic scientist rises or falls on the merit of his or her discovery because science allows for little ethical or intellectual ambiguity. Charismatic business leaders such as Lee Iacocca may also be warmly regarded, but in the commercial world it is more widely recognised that there have been some destructive individuals among them. Corporate leaders operate in freer environments than do scientists, but they are still hedged in by lawyers and accountants, shareholders and boards of directors, government regulations and the market. There is greater room for unethical behaviour, but still not a lot.

The evaluation of charismatic politicians is much more problematic; for every Mandela there has been a Mao. Politicians are freer than business leaders and much freer than scientists. Those arising in undemocratic countries may face little external restraint, and even those emerging from within democracy may rise above it, as Hitler did. Even in more robust democracies, there still may be ample opportunity for the politician to lie and cheat, and many inducements to do so.

The evaluation of charismatic religious leaders reveals total polarisation. The successful prophet becomes sanctified as God's messenger, but anything short of this sees him reviled as a cult leader. By definition the prophet is attempting to propound an ethic and practice that transcends human society, and is attempting to lead his followers to the Promised Land, either geographically or in their souls. Indeed, the prophet's great appeal is that he embodies something otherworldly, and appeals to divine law rather than to human conventions. However, a sacred mission may lead to secular temptations.

In sum, it is the degree of external restraint that their context forces upon them that largely determines how history evaluates charismatic personalities. If the leader faces strict ethical constraints, then he may have to resort to creativity of a very high order to achieve narcissistic supply, and this could bring out the best in him. Without such constraint, charismatic leadership sometimes results in disaster.

It is our species' fractiousness, but also our creativity that necessitates charisma. More than such specific evolutionary adaptations as language, the cortex, opposable thumbs, upright gait, social behaviour and extended gestation, it is our burgeoning creativity that underpins our adaptability. But such powerful creativity also causes problems in the forms of dissent and deviance, ambition and individualism. Under normal conditions these behaviours perform positive functions within society, and their extremes are held in check by traditional and rational authorities. But during times of crisis, stress or transition, such behaviours may threaten to fragment or undermine the social system. The role of charisma is to adhere and guide the group through difficult times, even sacrificing some if necessary, to bridge from one normative state to another with a minimum of loss and damage. Charisma is a high-risk problem-solving strategy that can restrain chaotic and destructive social behaviours, by force if necessary, while enabling the group to reach a new status quo wherein more functional traditional and rational authorities can re-assert themselves. The risk is that it may also unleash other destructive impulses. Charisma is an incipient potential, for good and evil, ever-present in the background and permeating all institutions. Contrary to Max Weber's theory, much of what passes for traditional and rational authority is actually dictated by the strongest personality on the block, and much of history follows a narrative line derived from charisma. We are all followers in our ways.

Weber addressed the evaluation of charisma in two late essays, 'Politics as a Vocation' and 'Science as a Vocation' (1946). He took a stand on classical democratic values (Bachrach, 1967), and this led him to reject charisma as being ultimately opposed to human growth, freedom and responsibility. However, he ended by challenging his students to engage with it in order to investigate, in the light of their own ultimate values, its meaning for them (Dow, 1978). In simple terms, this is to suggest that while at an institutional level charisma is antithetical to human progress, at a personal level — the level of the charismatic crucible — it can be a powerfully instructive experience.

In contrast, Charles Lindholm took a more relativistic stance. He evaluated charisma as 'empty', and having 'no substantive content' beyond a visceral experience; a judgment that renders it almost worthless (Lindholm, 1990, p. 164). But if this is all charisma is, then why does it grip people so, some of whom even sacrifice their lives for it? Rather, charisma offers an ecstatic encounter with our ultimate concerns, embodied in a living person. Although we may be unable — and perhaps unwise — to attempt to sustain this, probably no other experience (with the possible exceptions of romantic love and parenthood), can reveal us to ourselves so totally.

Precisely what is revealed? The answer to this question lies in the paradox of the charismatic condition, for perhaps only a narcissist can teach us so much about connection. As the scholars discussed herein have described, the charismatic personality uses others as parts of himself, and tries to solve for them some problem that he has been unable to solve for himself. He teaches best that which he most needs to learn. Charisma is perhaps the ultimate arena wherein a self–other interface can be explored and tested. The charismatic leader uses others to struggle with himself and the followers use him in their struggles with their selves.

Disillusionment is inevitable, for both participants. But in the electrifying blurring of the boundaries of self and other, in the hope and communion and lived vitality of charisma, the follower may at times rise above their troubled self and glimpse Dietrich Bonhoeffer's 'beyond in the midst of life'. What they do with that insight is, of course, up to them; charisma is a window, not a door. But if the follower learns well and succeeds in his great work, he may clarify the meaning of his life, gain profound access to his depths, knowledge of his extremes and an abiding conviction that no God or Other could possibly know him better than he now knows himself.

As for the leader, there are several possibilities. If he can make his false self work for him and achieve success in a good cause, then he may gain the kind of resolution that Churchill and Roosevelt found. There can be peace in this, and a sense of a battle well fought. More likely, however, he will fail. The challenge then becomes one of growing from his failure. Hopefully his failure comes early, before he becomes addicted to narcissistic supply, for even good men such as Churchill and Roosevelt clung to power far too long. Some refuse to accept defeat, returning again and again to the fray, and ending their days in bitterness. But a personality disorder is not a death sentence. If he is not too damaged, and if he has other creative interests that distract him from the easy seductions of power, typically family, or artistic or intellectual passions (or all three), then he may reflect upon and confront his narcissistic shortcomings. In time he may learn to connect more with others, perhaps indirectly or at a distance, and he may gain much satisfaction from this.

His narcissism remains and so does his interpersonal power, though he may prefer to withdraw into himself. There may be sudden lapses of rapport that confuse him. He may still be socially inappropriate at moments. He may be tempted at times to cut through the collective chatter with a well-timed charismatic barb. But to recover from such a severe disadvantage as a narcissistic personality disorder, to struggle through oneself to a deeper reality, takes courage and great effort, and the rare person who manages even a partial cure may, like Erik Erikson, achieve a painful wisdom. At best he may become a Sage.

His grandiosity remains also, but quieter now. For there are moments, perhaps sometimes in sunny slumbers, when he dreams that:

> Suddenly I am raised aloft by primordial passion;
> I become Leader, Law, Light, Prophet, Father, Author, and Journey,
> Rising above this world to the others that shine in splendor.
> I wander through every part of that ethereal country;
> Then, far away, as they gape at the marvel, I leave them behind me.
>
> *Giordano Bruno* (Rowland, 2008)

• • • • •

APPENDIX A

• · · · ·

Kernberg's Symptomatology of Narcissistic Personalities

Otto Kernberg's work on narcissistic personalities was presented in his 1975 book titled *Borderline Conditions and Pathological Narcissism* (New York, Jason Aronson). His characteristics include:

- An unusual degree of self-reference in their interactions with other people.
- An inflated self-concept.
- Extreme self-centredness.
- Idealisation of people from whom they expect narcissistic supplies (that is, applause and tokens of affection).
- Arrogance, being a defence against paranoid traits.
- A shallow emotional life.
- Little enjoyment from life itself.
- Boredom and restlessness.
- 'Emptiness behind the glitter'.
- Little empathy for the feelings of others.
- Control of others.
- Very dependent on others for tribute and praise, but unable really to depend on anyone because of a deep distrust.
- Strong conscious feelings of inferiority and insecurity (in a functional charismatic personality such feelings would be unconscious or strongly denied).
- A great need to be loved by others.
- Coldness and ruthlessness behind the charm.
- Envy of others.
- Superficially there is a lack of attachments to others, but actually 'their interactions reflect very primitive and intense (attachments) of a frightening kind'.
- Relations with people are exploitative and parasitic.
- Extreme contradictions such as fluctuating inferiority versus grandiosity, pursuit of power versus impotence.

REFERENCES

· · · · ·

Alderfer, C.P. (1972). *Existence, relatedness and growth: Human needs in organisational settings*. New York: Free Press.

Andreasen, N.C. (1987). Creativity and mental illness: Prevalence rates in writers and first-degree relatives. *American Journal of Psychiatry, 144,* 1288–1292.

American Psychiatric Association. (1994). *Diagnostic and statistical manual of mental disorders* (4th ed.). Washington: American Psychiatric Association.

Argyle, M. (2002). *The psychology of happiness*. London: Routledge.

Arrowsmith, W. (1958). Introduction to the Baccae. In: C. Grene & D. Lattimore (Eds.), *The complete Greek tragedies* (Vol. IV, pp. 537–538). Chicago: University of Chicago Press.

Athens, L. (1992). *The creation of dangerous violent criminals*. Urbana: University of Illinois Press.

Auden, W.H. (2007). *Selected poems*. New York: Vintage.

Aviolo, B.J., & Yammarino, F.J. (Eds.). (2002). *Transformational and charismatic leadership: The road ahead*. Oxford, UK: Elsevier.

Babiak, P., & Hare, R. (2006). *Snakes in suits: When psychopaths go to work*. London: Regan Books.

Bachrach, P. (1967). *The theory of democratic elitism: A critique*. Boston: Little, Brown and Company.

Baker, N. (2008). *Human smoke: The beginnings of World War II, the end of civilization*. New York: Simon and Schuster.

Balint, M. (1968). *The basic fault*. London: Hogarth Press.

Bandler, R., & Grinder, J. (1975). *Patterns of the hypnotic techniques of Milton Erickson*, (Vol. 1). Palo Alto, CA: Meta Publications.

Bandler, R., & Grinder, J. (1979). *Frogs into princes*. Moab, UT: Real People Press.

Barnes, D.F. (1978). Charisma and religious leadership: An historical analysis. *Journal for the Scientific Study of Religion, 17,* 1–18.

Bass, B.M. (1985). *Leadership and performance beyond expectations*. New York: Free Press.

Bass, B.M., & Riggio, R.E. (Eds.). (2006). *Transformational leadership*. Mahwah, N.J: Lawrence Erlbaum.

Batson, C.D., & Ventis, W.L. (1982). *The religious experience: A social psychological perspective*. Oxford, UK: Oxford University Press.

Baynes, N.H. (Ed.). (1942). *The speeches of Adolf Hitler*. New York: Oxford University Press.

Becker, E. (1962). *The birth and death of meaning: An interdisciplinary perspective on the problem of man*. New York: Free Press.

Becker, G. (2000–2001). The association of creativity and psychopathology: Its cultural–historical origins. *Creativity Research Journal, 13*, 45–53.

Behling, K. (2005). *Martha Freud: A biography*. Cambridge, UK: Polity.

Benjamin, L.S. (1996). *Interpersonal diagnosis and treatment of personality disorders*. New York: Guilford Press.

Berger, P.L. (1963). Charisma and religious innovation: The social location of Israelite prophecy. *American Sociological Review, 28*, 940–50.

Berman, L. (2003). *The puzzle: Exploring the evolutionary puzzle of male homosexuality*. Pittsburgh, PA: Godot Press.

Biever, C. (2009, Jaury 3). The makings of a savant. *New Scientist*, pp. 40–41.

Bishop, J. (1975). *FDR's last year*. London: Harte-Davis, MacGibbon.

Binion, R. (1991). *Hitler among the Germans*. DeKalb, IL: Northern Illinois University Press.

Blau, P.M. (1963). Critical remarks on Weber's theory of authority. *American Political Science Review, 57*, 305–16.

Bloland, S.E. (2005). *In the shadow of fame: A memoir by the daughter of Erik H. Erikson*. New York: Viking.

Bloom, V. (1998, May). *Psychoanalysis and creativity*. Paper presented at the Spring meeting of the American Association of Psychoanalysis, Toronto, Canada.

Bohart, A.C., & Tallman, K. (1999). *How clients make therapy work: The process of active self-healing*. Washington DC: American Psychological Association.

Bonhoeffer, D. (1995). *A testament to freedom: The essential writings of Dietrich Bonhoeffer*. New York: HarperOne.

Bonhoeffer, D. (1997). *Letters and papers from prison*. Carmichael, CA: Touchstone.

Boring, E.G. (1961). *Psychologist at large*. New York: Basic Books.

Bradley, R.T. (1987). *Charisma and social structure: A study of love and power, wholeness and transformation*. New York: Paragon House.

Breger, L. (2000). *Freud: Darkness in the midst of vision*. New York: John Wiley.

Brener, M.E. (2005). *Richard Wagner and the Jews*. New York: McFarland and Company.

Brenner, L. (2007, March 21). *Interview: The Religion Report.* Radio National, Australian Broadcasting Commission.

Britton, R. (1998). *Belief and imagination.* London: Routledge.

Brod, J.H. (1997). Creativity and schizotypy. In G. Claridge (Ed.), *Schizotypy: Implications for illness and health* (pp. 274–298). Oxford: Oxford University Press.

Brown, F.W., & Moshavi, D. (2002) .Herding academic cats: Faculty reactions to transformational and contingent reward leadership by departmental chairs. *Journal of Leadership Studies, 8,* 79–94.

Brown, F.W., & Trevino, L.K. (2003, August). *The influence of leadership styles on unethical conduct in work groups: An empirical test.* Paper presented at the meeting of the Academy of Management, Seattle, WA.

Burke, K.L., & Brinkerhoff, M.B. (1981). Capturing charisma: Notes on an elusive concept. *Journal for the Scientific Study of Religion, 20,* 274–284.

Burnham, K.E. (1979). *God comes to America: Father Divine and the Peace Mission Movement.* New York: Lambeth Press.

Burns, J.M. (1978). *Leadership.* New York: Harper and Row.

Camic, C. (1980). Charisma: Its varieties, preconditions and consequences. *Sociological Inquiry, 50,* 5–23.

Carlyle, T. (2007). *On heroes and hero worship and the heroic in history.* Charleston, SC: Bibliobazaar.

Carter, V.B. (1965). *Winston Churchill: As I knew him.* London: Reprint Society.

Chafe, W.H. (2005). *Private lives, public consequences: Personality and politics in modern America.* London: Harvard University Press.

Chait, J. (2009, March 18). Wasting away in Hooverville. *The New Republic,* pp. 38–40.

Chang, J., & Halliday, J. (2005). *Mao: The unknown story.* London: Jonathan Cape.

Choderow, N. (1999). *The reproduction of mothering: Psychoanalysis and the psychology of gender.* Los Angeles: University of California Press.

Churchill, R.S. (1966). *Winston S. Churchill* (Vols. 1–3). London: Heinemann.

Churchill, W.S. (1930). *My early life.* London: Odhams.

Clare, A. (1997, November 16). That shrinking feeling. *The Sunday Times,* pp. 8–10.

Clark, R. (1980). *Freud: The man and the cause.* London: Ballantine.

Cohen, L. (1974). Field Commander Cohen. On *New Skin for the Old Ceremony.* New York: Columbia Records.

Coles, R.W. (1973). *A sociological yearbook of religion in Britain.* London: SCM Press.

Conger, J.A., & Kanungo, R.N. (1988). *Charismatic leadership: The elusive factor in organizational effectiveness.* London: Jossey-Bass.

Conrad, J. (1902, 1950). *Heart of darkness and the secret sharer.* New York: Signet.

Csikszentmihalyi, M. (1991). *Flow: The psychology of optimal experience.* New York: Harper Perennial.

Davis, K.S. (1984, Winter). FDR as a biographer's problem. *The American Scholar,* pp. 83–88.

Dawes, R.M. (1994). *House of cards: Psychology and psychotherapy built on myth.* New York: Free Press.

De Bono, E. (1970). *Lateral thinking: A textbook of creativity.* London: McGraw Hill.

DePaulo, B.M., & Friedman, H.S. (1998). Nonverbal communication. In D.T. Gilbert, S.T. Fiske, & G. Lindzey (Eds.), *The handbook of social psychology* (Vol. 2, 4th ed., pp. 3–40). New York: McGraw Hill.

Donaldson, M. (2003). Studying up: The masculinity of the hegemonic. In S. Tomsen & M. Donaldson, *Male trouble: Looking at Australian masculinities* (pp. 156–179). Melbourne, Australia: Pluto Press.

Dorfman, P.W., Hanges, P.J., & Brodbeck, F.C. (2004). Leadership and cultural variation: The identification of culturally endorsed leadership profile. In R.J. House, P.J. Hanges, M. Javidan, P.W. Dorfman, & V. Gupta (Eds.), *Culture, leadership and organizations: The GLOBE study of 62 societies* (pp. 669–719). Thousand Oaks, CA: Sage.

Dow, T.E. (1978). An analysis of Weber's work on charisma. *British Journal of Sociology, 29,* 83–93.

Downs, A. (2005). *The velvet rage: How growing up gay in a straight man's world can lead to destructive anger or a creative edge.* Cambridge, MA: Da Capo Lifelong Books.

Dumdum, U.R., Lowe, K.B., & Avolio, B.J. (2002). A meta-analysis of transformational and transactional leadership correlates of effectiveness and satisfaction: An update and extension. In B.J. Avolio & F.J. Yammarino (Eds.), *Transformational and charismatic leadership: The road ahead* (pp. 35–66). Oxford, UK: Elsevier.

Dutton, D.G. (1997). *The batterer: A psychological profile.* New York: Basic.

Dutton, D. (2006). *Rethinking domestic violence.* Vancouver, Canada: University of British Columbia Press.

Dutton, D.G. (2007). *The abusive personality: Violence and control in intimate relationships* (2nd ed.). Boston: Guilford.

Eagly, A.H., Johannesen-Schmidt, M.C., & van Eigen, M.L. (2003). Transformational, transactional and laissez-faire leadership styles: A meta-analysis comparing men and women. *Psychological Bulletin, 129*, 569–591.

Eden, D., & Sulimani, R. (2002). Pygmalion training made effective: Greater mastery through augmentation of self-efficacy and means efficacy. In B.J. Avolio & F.J. Yammarino (Eds.), *Transformational and charismatic leadership: The road ahead* (pp. 287–308). Oxford, UK: Elsevier.

Eliade, M. (1964). *Shamanism: Archaic techniques of ecstasy* (W.R. Trask, Trans.). Princeton, NJ: Princeton University Press.

Elkins, T., & Keller, R.T. (2003). Leadership in research and development organisations: A literature review and conceptual framework. *The Leadership Quarterly, 14*, 587–606.

Ellenberger, H. (1970). *The discovery of the unconscious: The history and evolution of dynamic psychiatry.* New York: Basic Books.

Ellis, A. (1996). How I learned to help clients feel better and get better. *Psychotherapy, 33*, 149–151

Ellis, E. (2007, June). Wendi Deng Murdoch. *The Monthly*, 28–40.

Emmons, N. (1988). *Manson in his own words.* New York: Grove Press.

Ericcson, K.A. (2006). *Cambridge handbook of expertise and expert performance.* Cambridge, UK: Cambridge University Press.

Ericsson, K.K., Chase, W.G., & Faloon, S. (1980) Acquisition of a memory skill. *Science, 208*, 1181–1182

Erikson, E.H. (1942). Hitler's imagery and German youth. *Psychiatry, 5*, 475–493.

Erikson, E.H. (1969). *Gandhi's truth: On the origins of militant non-violence.* New York: Norton.

Fairbairn, R. (1952). *Collected papers.* London: Hogarth Press.

Fallowell, D. (1988, January 24). Greer: A woman of substance. *Times On Sunday*, pp. 61–62.

Farber, L.H. (1961, Fall). Will and willfulness in hysteria. *Review of Existential Psychology and Psychiatry*, 165–182.

Farber, L.H. (2000). *The ways of the will: Selected essays.* New York: Basic Books.

Fenichel, O. (1946). On acting. *The Psychoanalytic Quarterly, 15*, 148–156.

Ferguson, S. (2003, December 6). Selling the dream, buying the nightmare. *The Age* [Insight], p. 9.

Feyerabend, P. (1975). *Against method: Outline of an anarchistic theory of knowledge.* London: Verso.

Fillion, K. (1997). *Lip service: The truth about women's darker side in love, sex and friendship.* Toronto, Canada: Harper Collins.

Fischer, L. (1951). *The life of Mahatma Gandhi*. London: Harper Collins.

Flexner, J.T. (1974). *Washington: The indispensable man*. Boston: Little, Brown and Company.

Forster, E.M. (1995). *Howard's end*. Harmondsworth, UK: Penguin. (Original work published 1910).

Foster, L. (1983). A personal odyssey: My encounter with Mormon history. *Dialogue, 16*(3), 87–98.

Foster, L. (1984). Career apostates: Reflections on the works of Jerald and Sandra Tanner. *Dialogue, 17*, 35–60.

Freud, S. (1914). On narcissism. In *The standard edition of the complete psychological works of Sigmund Freud* (Vol. XIV, pp. 73–102). (James Strachey, Trans). London: Hogarth Press.

Freud, S. (1916). Some character types met with in psychoanalytic work. In *The standard edition of the complete psychological works of Sigmund Freud* (Vol. XIV pp. 309–333). (James Strachey, Trans). London: Hogarth Press.

Freud, S. (1921, 1960) *Group psychology and the analysis of the ego* (James Strachey (Trans). New York: Bantam.

Freud, S. (1937/2001). Analysis terminable and interminable. In *The Standard Edition of the Complete Psychological Works of Sigmund Freud* (Vol. XXIII). (James Strachey, Trans.). London: Hogarth Press.

Fromm, E. (1964). *The heart of man: Its genius for good and evil*. New York: Harper.

Gaines, J. (1979). *Fritz Perls: Here and now*. Millbrae, CA: Celestial Arts.

Gallese, V. (2001). The 'shared manifold' hypothesis: From mirror neurones to empathy. *Journal of Consciousness Studies, 8*, 87–96.

Gandhi, M.K. (1927). *An autobiography, or the story of my experiments with truth*. Harmondsworth, UK: Penguin.

Gartner, J.D. (2005). *The hypomanic edge: The link between (a little) craziness and (a lot of) success in America*. New York: Simon and Schuster.

Gilbert, G.M. (1950). *The psychology of dictatorship*. New York: Ronald Press.

Goldhagen, D.J. (1996). *Hitler's willing executioners: Ordinary Germans and the Holocaust*. New York: Knopf.

Goleman, D. (1996). *Emotional intelligence: Why it can matter more than IQ*. London: Bloomsbury.

Gordon, J.S. (1987). *The golden guru: The strange journey of Bhagwan Shree Rajneesh*. Lexington, MA.: Stephen Greene.

Greer, G. (1989). *Daddy we hardly knew you*. London: Hamish Hamilton.

Greer, G. (1991). *The female eunuch*. London: Paladin.

Greer, G. (2003). *The beautiful boy*. New York: Rizzoli.

Grenier, R. (1983). *The Gandhi nobody knows*. Nashville, TN: Thomas Nelson.

Grosskurth, P. (1995). *Melanie Klein: Her world and her work*. Northvale, NJ.: Jason Aronson.

Guiloineau, J. (2002). *The early life of Rolihlahla Madiba Nelson Mandela*. Berkeley, CA: North Atlantic Books.

Gunderson, J.G., Weinberg, I., Daversa, M.T., Kueppenbender, K.D., Zanarini, M.C., Shea, T., Skodol, A.E., Sanislow, C.A., Yen, S., Morey, L.C., Grilo, C.M., McGlashan, T.H. Stout, R.L., & Dyck, I. (2006). Descriptive and longitudinal observations on the relationship of borderline personality disorder and bipolar disorder. *American Journal of Psychiatry, 163*, 1173–1178.

Haber, R.N., & Haber, R.B. (1964) Eidetic imagery I: Frequency. *Perceptual and Motor Skills, 19*, 131–138.

Hackman, M.Z., Furniss, A.H., Hills, M.J., & Paterson, R.J. (1992). Perceptions of gender role characteristics and transformational and transactional leadership behaviours. *Perceptual and Motor Skills, 75*, 311–319.

Haffner, S. (1979). *The meaning of Hitler*. London: Phoenix.

Haidt, J. (2008). What makes people vote Republican? *Edge: The Third Culture*. www.edge.org/3rd_culture/haidt08/haidt08_index.html

Haley, J. (1973). *Uncommon therapy*. New York: Grune and Stratton.

Hall, L.K. (1983). *Charisma: A study of personality characteristics of charismatic leaders*. Unpublished doctoral dissertation, University of Georgia.

Harrison, B.G. (1992, September). Germaine Greer: After the change. Mirabella, p. 88.

Heiber, H. (Ed.) (1962). *The early Goebbels diaries: 1925–1926*. New York: Praeger.

Heidel, A. (Ed., Trans.) (1968). The Gilgamesh Epic. In W.H. McNeill & J.W. Sedlar (Eds.), *The origins of civilisation*. London: Oxford University Press.

Hellinger, B. (1998). *Love's hidden symmetry: What makes love work in relationships*. Redding, CT: Zeig, Tucker and Theisen

Hendricks, K. (2004). *The relationship between creativity dimensions of normal personality and scales of personality disorder among artists and non-artists*. Unpublished doctoral dissertation, University of Melbourne, Australia.

Hennig, N., & Jardim, A. (1977). *The managerial woman*. Garden City, NY: Doubleday.

Hershman, D.J., & Lieb, J. (1988). *The key to genius: Manic depression and the creative life*. London: Prometheus.

Hershman, D.J., & Lieb, J. (1994). *A brotherhood of tyrants: Manic depression and absolute power.* London: Prometheus

Hershman, D.J., & Lieb, J. (1998). *Manic depression and creativity.* London: Prometheus.

Hesse, J. (1970, August 31). From champion majorette to Frank Sinatra date. *Vancouver Sun.*

Higham, C. (2006). *Dark lady: Winston Churchill's mother and her world.* London: Virgin.

Hitchens, C. (1977). *The missionary position: Mother Teresa in theory and practice.* New York: Verso.

Hitler, A. (1939). *Mein Kampf* (English edition). New York: Reynal and Hitchcock.

Hogan, R., Raskin, R., & Fazzini, D. (1990). The dark side of charisma. In K.E. Clark & M.B. Clark (Eds.), *Measures of leadership*. West Orange, NJ.: Leadership Library of America.

Holmes, J. (2001). *Narcissism.* Cambridge: Icon Books.

Horney, K. (1937). *The neurotic personality of our time.* New York: W.W. Norton.

Horney, K. (1945). *Our inner conflicts.* New York: Norton.

Houselander, C. (1952). *Guilt.* London: Sheed and Ward.

Hoyt, C.L., & Blascovitch, J. (2003). Transformational and transactional leadership in virtual and physical environments. *Small Group Research, 34,* 678–715.

Hughes, T. (1997). *Tales from Ovid.* London: Faber and Faber.

Hume, J.C. (1994). *The wit and wisdom of Winston Churchill.* New York: HarperCollins.

Hunt, K. (2007, October 6). Untamed shrew is not convincing. *Weekend Australian Review,* p. 13).

Iacocca, L. (1984). *Iacocca: An autobiography* (with William Novak). New York: Bantam.

James, C. (2008). *Cultural amnesia: Necessary memories from history and the arts.* New York: Norton.

James, W. (1961). *Varieties of religious experience.* New York: Collier MacMillan.

Jamison, K.R. (1993). *Touched with fire: Manic-depressive illness and the artistic temperament.* New York: Free Press.

Janda, L. (1996). *The psychologist's book of self-tests.* New York: Perigree.

Jenkins, R. (2001). *Churchill.* London: MacMillan.

Jenkins, R. (2003). *Franklin Delano Roosevelt.* London: Pan.

Jones, E. (1955). *The life and work of Sigmund Freud* (Vols. 1–3). New York: Basic Books.

Jung, C.G. (1954). *The development of personality. The collected works of C.G. Jung* (Vol. 17), (R.F.C. Hull, Trans.). London: Routledge and Kegan Paul.

Jung, C.G. (1959). *Aion. The collected works of C.G. Jung* (Vol. 9), Edited by H. Read, M. Fordham, & G. Adler. Bollingen Series. New York: Pantheon Books.

Jung, D.I. (2001). Transformational and transactional leadership and their effects on creativity in groups. *Creativity Research Journal, 13*, 185–195.

Junge, T. (2003). *Until the final hours: Hitler's last secretary.* London: Phoenix.

Kahai, S.S., Sosik, J.J., & Avolio, B.J. (2003). Effects of leadership style, anonymity, and rewards on creativity-relevant processes and outcomes in an electronic meeting system context. *The Leadership Quarterly, 14*, 499–524.

Kaplan, L.J. (1979). *Oneness and separateness: From infant to individual.* London: Jonathan Cape.

Kazantzakis, N. (1971). *Zorba the Greek.* New York: Touchstone (Simon and Schuster).

Kephart, W.M. (1982). *Extraordinary groups.* New York: St Martins Press.

Kernberg, O. (1975). *Borderline conditions and pathological narcissism.* New York: Jason Aronson.

Kerr, M., & Bowen, M. (1988). *Family evaluation.* New York: Norton.

Kershaw, I. (2000). *Hitler: 1989–1936, Hubris.* New York: Norton.

Kershaw, I. (2001). *Hitler: 1936–1945, Nemesis.* New York: Norton.

Kershaw, I. (2007). *Fateful choices: Ten decisions that changed the world, 1940–1941.* Harmondsworth, UK: Penguin.

Khoo, H.S., & Burch, G.St.J. (2008). The 'dark side' of leadership personality and transformational leadership: An exploratory study. *Personality and Individual Difference, 44*(1), 86–97.

Kiesler, D.J. (1996). *Contemporary interpersonal theory and research: Personality, psychopathology and psychotherapy.* New York: Wiley.

Kilbourne, B.K., & Richardson, J.T. (1980) People's Temple and Jonestown: A corrective comparison and critique. *Journal for the Scientific Study of Religion, 19*(3) 239–255.

Kluckhohn, C., & Murray, H.A. (1953). Personality formation: The determinants. In C. Kluckhohn, H. Murray, & D. Schneider (Eds.), *Personality in nature, society and culture.* New York: Knopf.

Kohut, H. (1966). Forms and transformations of narcissism. *Journal of the American Psychoanalytic Association, 14*, 243–272.

Kohut, H. (1971). *The analysis of the self: A systematic approach to the psychoanalytic treatment of narcissistic disorders.* New York: International Universities Press.

Kohut, H. (1972). Thoughts on narcissism and narcissistic rage. *Psychoanalytic Study of the Child, 27,* 360–400.

Kohut, H. (1976). Creativeness, charisma and group psychology: Reflections on the self-analysis of Freud. In J.E. Gedo & G.H. Pollock (Eds.), *Freud: The fusion of science and humanism; The intellectual history of psychoanalysis.* New York: International Universities Press.

Kohut, H. (1977). *The restoration of the self.* New York: International Universities Press.

Kohut, H. (1980). Reflection on advances in self-psychology. In A. Goldberg, (Ed.), *Advances in self-psychology.* New York: International Universities Press.

Kohut, H. (1985). *Self-psychology and the humanities: Reflections on a new psychoanalytic approach.* Edited by C.B. Strozier. New York: Norton.

Kopp, S.B. (1971). *Guru: Metaphors from a psychotherapist.* Palo Alto, CA: Bantam.

Kopp, S.B. (1972). *If you meet the Buddha on the road, kill him!* New York: Bantam.

Korenberg, J. (2009, February 21). Rare syndrome sheds light on sociability. *New Scientist,* p. 14.

Kottler, J.A. (2006). *Divine madness: Ten stories of creative struggle.* San Francisco. Jossey-Bass.

Kuhn, T. (1970). *The structure of scientific revolutions* (2nd ed.). Chicago: University of Chicago Press.

Labak, A.S. (1972). *The study of charismatic college teachers.* Unpublished doctoral dissertation, University of Northern Colarado.

La Barre, W. (1980). *Culture in context: Selected writings of Weston La Barre.* Durham, NC: Duke University Press.

Langton, M. (2008, August 19). Greer maintains the rage of racists. *The Australian,* p. 20.

Le Bon, G. (2006). *The crowd: A study of the popular mind.* London: Filiquarian Publishing, LLC.

Levine, S.V. (1984). *Radical departures: Desperate detours to growing up.* New York: Harcourt Brace Jovanovich.

Lewis, D. (2003). *The man who invented Hitler.* London: Headline.

Li, Z. (1996). *The private life of Chairman Mao* (T. Hung-chao, Trans.) London: Arrow.

Lindholm, C. (1990). *Charisma*. Cambridge, MA: Basil Blackwell.

Little, G. (1985). *Political ensembles: A psycho-social approach to politics and leadership*. Melbourne, Australia: Oxford University Press.

Livesley, W.J., Jackson, D.N., & Schroeder, M.L. (1992). Factorial structure of traits delineating personality disorders in clinical and general population samples. *Journal of Abnormal Psychology, 101*, 432–440.

Lodge, T. (2006). *Mandela: A critical life*. Oxford: Oxford University Press.

Loewenstein, K. (1966). *Max Weber's political ideas in the perspective of our time*. Amherst, MA: University of Massachusetts Press.

Lowe, K.B., Kroeck, K.G., & Sivasubramianiam, N. (1996). Effectiveness correlates of transformational and transactional leadership: A meta-analytic review of the MLQ literature. *The Leadership Quarterly, 7*, 385–425.

Ludwig, A.M. (1995). The price of greatness. *Journal of Abnormal Psychology, 110*, 401–412.

Lukacs, J. (1997). *The Hitler of history*. New York: Knopf.

Lumley, C. (2005, January 14). Germaine Greer has become the Michael Jackson of feminism. *The Age*, p. 24.

Lynn, D.J. (1997). Sigmund Freud's psychoanalysis of Albert Hirst. *Bulletin of the History of Medicine, 71*, 69–93.

Machiavelli, N. (1950). *The prince and the discourses*. New York: McGraw-Hill.

Machtan, L. (2001). *The hidden Hitler*. New York: Basic.

MacMillan, M. (1991). *Freud evaluated: The completed arc*. Amsterdam: New Holland.

Magai, C., & Haviland-Jones. (2002). *The hidden genius of emotion: Lifespan transformations of personality*. Cambridge: Cambridge University Press.

Mahler, M., Pine, F., & Bergman, A. (1967). *The psychological birth of the human infant*. New York: Basic Books.

Malcolm, J. (1980). *Psychoanalysis: The impossible profession*. London: Pan.

Malcolm, N. (2009, February 26). The guru of gossip. Review of *Maurice Bowra: A Life,* by Leslie Mitchell (2008). *Standpoint Magazine*, pp. 41–42.

Malle, L., (Producer and Director). (1992). *Damage* [Motion picture]. New York: New Line Cinema.

Mandela, N. (1994). *Long walk to freedom*. London: Abacus.

Mann, G. (2000, October). *Edges of mind*. Paper presented to the Second Annual Conference of the Tel Aviv Institute for Contemporary Psychoanalysis, Tel Aviv.

Martines, L. (2007). *Scourge and fire: Savonarola in renaissance Italy*. London: Pimlico.

Maslow, A. (1954). *Motivation and personality*. New York: Harper and Row.

Maslow, A. (1968). *Towards a psychology of being*. New York: Wiley.

Masson, J. (1984). *The assault on truth: Freud's suppression of the seduction hypothesis*. New York: Farrar, Strauss and Giroux.

Masson, J. (1985). *The complete letters of Sigmund Freud to Wilhelm Fliess 1887–1904*. Boston: Harvard University Press.

Masson, J. (1990). *Against therapy: Emotional tyranny and the myth of psychological healing*. London: Fontana.

Masterson, J.F. (1988). *The search for the real self: Unmasking the personality disorders of our age*. New York: Free Press.

May, R. (1973). *Man's search for himself*. New York: Delta.

May, R. (1973). *Paulus*. New York: Harper and Row.

May, R. (1975). *The courage to create*. New York: Norton.

McColl-Kennedy, J.R., & Anderson, R.D. (2002). Impact of leadership style and emotions on subordinate performance. *The Leadership Quarterly, 13*, 545–559.

McFarlin, D., & Sweeney, P. (2002). *Where egos dare: The untold truth about narcissistic leaders and how to survive them*. London: Kogan Page.

McHugh, P.R. (2006). *The mind has mountains: Reflections on society and psychiatry*. Baltimore: Johns Hopkins University Press.

McWilliams, N. (1994). *Psychoanalytic diagnosis*. New York: Guilford.

Meindl, J.R. (1995). The romance of leadership as a follower-centric theory: A social constructionist approach. *Leadership Quarterly, 6*, 329–341.

Millon, T. (1996). *Disorders of personality: DSM IV and beyond*. New York: Wiley.

Millon, T., Millon, M.C., Meagher, S., Grossman, S., & Ramnath, R. (2004). *Personality disorders in modern life*. New York: Wiley.

Milne, H. (1986). *Bhagwan: The god that failed*. Edited by Liz Hodgkinson. London: Caliban.

Mitchell, L. (2008). London: Oxford University Press.

Mollon, P. (1993). *The fragile self*. London: Whurr.

Moore, S. (2001). Bursting the bubble: The flip side of love in charismatic groups; Love, hate and a consideration of the death drive. *Australian Journal of Psychotherapy, 19*, 77–94.

Murphy, M. (1992). *The future of the body: Explorations into the further evolution of human nature*. Los Angeles: Jeremy Tarcher.

Oakes, L.D. (1997). *Prophetic charisma: The psychology of revolutionary religious personalities*. New York: Syracuse University Press.

Oakes, L.D. (1986). *Inside Centrepoint: The story of a New Zealand community.* Auckland, New Zealand: Benton Ross.

Ochse, R. (1991). The relation between creative genius and psychopathology: An historical perspective. *South African Journal of Psychology, 21,* 45–53.

Olden, C. (1941). About the fascinating effect of the narcissistic personality. *American Imago, 2,* 347–355.

Olin. W.F. (1980). *Escape from Utopia: My ten years in Synanon.* Santa Cruz, CA: Unity Press.

Olin, S.C. (1980). The Oneida Community and the instability of charismatic authority. *Journal of American History, 67,* 285–300.

Olivier, L. (1984). *Confessions of an actor.* New York: Penguin.

Orion, D. (1977). *I know you really love me.* New York: Macmillan.

Ostrom, J.W. (Ed.). (1948). *The letters of Edgar A. Poe* (Vol. 2). Cambridge, MA: Harvard University Press.

Patton, T. (1980). Foreword. In W. Olin, *Escape from utopia: My ten years in Synanon.* Santa Cruz: Unity Press.

Pells, R. (2008, May 21). The philistine's past. *The Australian* [Higher Education], p. 30.

Perls, F.S., Hefferline, R.F., & Goodman, P. (1951). *Gestalt Therapy: Excitement and growth in the human personality.* London: Souvenir.

Phillips, H. (2008, October 1). The outer limits of the human brain. *New Scientist,* pp. 28–33.

Pirsig, L. (2006). *Zen and the art of motorcycle maintenance.* New York: HarperTorch.

Polanyi, M. (1964). *Personal knowledge: Towards a post-critical philosophy.* New York: Harper and Row.

Post, F. (1994). Creativity and psychopathology: A study of 291 world-famous men. *British Journal of Psychiatry, 165,* 22–34.

Post, J.M. (2005). *The psychological assessment of political leaders: With profiles of Saddam Hussein and Bill Clinton.* Ann Arbor, MI: University of Michigan Press.

Ramsey, A., Watson, P.J., Biderman, M.D., & Reeves, A.L. (1996). Self-reported narcissism and perceived parental permissiveness and authoritarianism. *Journal of Genetic Psychology, 157,* 227–238.

Rawlinson, A. (1997). *The book of enlightened masters.* Chicago: Open Court.

Redlich, F. (1998). *Hitler: Diagnosis of a destructive prophet.* New York: Oxford University Press.

Richardson, J.T. (1995) Clinical and personality assessment of participants in new religions. *International Journal for the Psychology of Religion, 5*(3) 145–170.

Reich, W. (1963). *The murder of Christ.* New York: Farrar, Straus & Giroux.

Ride, C. (2002). *Narcissus and Echo.* Melbourne, Australia: Psychoz Publications.

Rienzi, B.M., Forquera, J., & Hitchcock D.L. (1995). Gender stereotypes for proposed DSM IV negativistic, depressive, narcissistic and dependent personality disorders. *Journal of Personality Disorders, 9,* 49–55.

Riggio, R.E. (1992). Social interaction skills and nonverbal behaviour. In R.S. Feldman (Ed.), *Applications of nonverbal behaviour theories and research* (pp. 3–30). Hillsdale, NJ: Lawrence Erlbaum Associates.

Roach, J. (2007). *It.* Minnesota: University of Michigan Press.

Robbins, T., & Anthony, D. (1995). Sects and violence. In: S. Wright (Ed.), *Armageddon in Waco* (pp. 236–259). Chicago: University of Chicago Press.

Roberts, B.W., & Mroczek, D. (2008). Personality trait change in adulthood. *Current Directions in Psychological Science, 17,* 31–35.

Rogers, C. (1959). A theory of therapy, personality and interpersonal relationships as developed in the client-centered framework. In S. Koch (Ed.), *Psychology: A Study of a Science : Vol. 3. Formulations of the Person and the Social Context* (pp. 184–256). New York: McGraw Hill.

Rogers, C. (1995). *On becoming a person: A therapist's view of psychotherapy.* San Francisco: Mariner.

Rowland, I. (2008). *Giordano Bruno: Philosopher/heretic.* New York: Farrar Straus and Giroux.

Rycroft, C. (1968). *A critical dictionary of psychoanalysis.* Harmondsworth, UK: Penguin.

Sampson, A. (1999). *Mandela: The authorised biography.* London: HarperCollins.

Samuels, A., Shorter, B., & Plaut, F. (1986). *A critical dictionary of Jungian analysis.* London: Routledge and Kegan Paul.

Satir, V. (1976). *Making contact.* Berkeley, CA: Celestial Arts.

Schmanlenbach, H. (1961). Communion: A sociological category (K. Naegele, Trans.) In T. Parsons, E. Shils, K. Naegele & J. Pitts (Eds.), *Theories of Society.* Glencoe, Ill.: Free Press.

Schweitzer, A. (1984). *The age of charisma.* Chicago: Nelson-Hall.

Semple, D., Smyth, R., Burns, J., Darjee, R., & McIntosh, A. (2005). *Oxford handbook of psychiatry.* Oxford, UK: Oxford University Press.

Sennett, R. (1975). Charismatic de-legitimation: A case study. *Theory and Society, 2,* 171–81.

Settle, J.E., Dawes, C.T., & Fowler, J.H. (2009). The heritability of partisan attachment. *Political Research Quarterly, 62*(3), 601–613.

Shamir, B. (2007). From passive recipients to active co-producers. In B. Shamir R. Pillai, M.C. Bligh, & M. Uhl-Bien (Eds.), *Follower-centred perspectives on leadership* (pp. 187-210). Greenwich, CT: Information Age Publishing.

Sharaf, M. (1983). *Fury on Earth: A biography of Wilhelm Reich*. London: Hutchinson.

Shepard, M. (1976). *Fritz*. New York: Bantam.

Short, P. (1999). *Mao: A life*. London: Hodder and Stoughton.

Silverman, L.H., & Weinberger, J. (1968). Mommy and I are one: Implications for psychotherapy. *American Psychologist, 40*, 1296–1308.

Smith, R., & Citrin, J.M. (2003). *The five patterns of extraordinary careers: The guide for achieving success and satisfaction*. New York: Crown Business (Random House).

Smith, W.C. (1962). *The meaning and end of religion*. London: SPCK

Smith, W.C. (1979). *Faith and belief.* Princeton, NJ: Princeton University Press.

Stern, D. (1985). *The interpersonal world of the infant*. New York: Basic Books.

Stevens, A. (1998). *An intelligent person's guide to psychotherapy*. London: Duckworth.

Stevens, B. (1970). *Don't push the river*. Moab, UT: Real People Press.

Stevens, A., & Price, J. (1996). *Evolutionary psychiatry: A new beginning*. London: Routledge.

Stierlin, H. (1959). The adaptation to the stronger person's personality. *Psychiatry, 22,* 143–152

Stone, M.H. (1993). *Abnormalities of personality: Within and beyond the realm of treatment*. New York: Norton.

Strozier, C.B. (1980). Heinz Kohut and the historical imagination. In A. Goldberg (Ed.), *Advances in self psychology* (pp. 1–28). New York: International Universities Press.

Strozier, C.B. (2001). *Heinz Kohut: The making of a psychoanalyst*. New York: Farrar, Strauss, Giroux.

Strozier, C.B., & Offer, D. (1985). *The leader: Psychohistorical essays*. New York: Springer.

Sulloway, F. (1979). *Freud, biologist of the mind: Beyond the psychoanalytic legend*. London: Andre Deutsche.

Summers, F. (1994). *Object relations theories and psychopathology: A comprehensive text*. Hillsdale, NJ: Analytic Press.

Symington, N. (1993). *Narcissism: A new theory*. London: Karnac.

Symington, N. (2002). *A pattern of madness*. London: Karnac.

Szasz, T. (1963). Freud's leadership. *Antioch Review, 22,* 153.

Tammet, D. (2009). *Embracing the wide sky: A tour across the horizons of the human mind.* London: Hodder and Staughton.

Thapar, A., & McGuffin, P. (1993). Is personality disorder inherited? An overview of the evidence. *Journal of Psychopathology and Behavioural Assessment, 15,* 325–345.

Tolstoy, L. (1986). *War and peace.* Harmondsworth, UK: Penguin.

Tomkins, S. (1987). Script theory. In J. Aronoff, A.I. Rabin, & R.A. Zucker (Eds.), *The emergence of personality* (pp. 147–216). New York: Springer.

Tomsen, S., & Donaldson, M. (Eds). (2003). *Male trouble: Looking at Australian masculinities.* Unpublished manuscript.

Troyat, H. (1994). *Catherine the Great* (J. Pinkham, Trans). New York: Meridian.

Trungpa, C. (1973). Cutting through spiritual materialism. Boston, MA: Shambhala.

Tschanz, B.T., Morf, C.C., & Turner, C.W. (1998). Gender differences in the structure of narcissism: A multi-sample analysis of the Narcissistic Personality Inventory. *Sex Roles, 38,* 863–870.

Tucker, R.C. (1968). The theory of charismatic leadership. *Daedelus, 97,* 731–756.

Turner, V.W. (1969). *The ritual process: Structure and anti-structure.* Chicago: Aldine.

Turner, V.W. (1974). *Dramas, fields and metaphors: Symbolic action in human society.* Ithaca, NY: Cornell University Press.

Vaillant, G.E. (2002). *Aging well.* Boston: Little, Brown and Company.

Vandenberghe, C., Stordeur, S., & D'hoore, W. (2002). Transactional and transformational leadership in nursing: Structural validity and substantive relationships. *European Journal of Psychological Assessment, 18,* 16–29.

Van der Braak, A. (2003). *Enlightenment blues: My years with an American guru.* Boston: Monkfish.

Van Vugt, M. (2008, June 14). Follow me. *New Scientist,* 42–45.

Waite, R.G.L. (1977). *The psychopathic God: Adolf Hitler.* New York: Signet.

Wallace, C. (1997). *Greer: Untamed shrew.* Sydney, Australia: MacMillan.

Waller, J. (2002). *Becoming evil: How ordinary people commit genocide and mass killing.* New York: Oxford.

Waller, N.G., Kojetin, B.A., Bouchard, T.J., Lykken, D.T., & Tellegen, A. (1990). Genetic and environmental influences on religious interests, attitudes and values: A study of twins reared apart and together. *Psychological Science, 1,* 138–142.

Wallis, W.D. (1943). *Messiahs: Their role in civilisation.* Washington, DC: American Council of Public Affairs.

Walter, J.A. (1985). Achievement and shortfall in the narcissistic leader: Gough Whitlam and Australian politics. In C.B. Strozier & D. Offer (Eds.), *The leader: Psychohistorical essays*. New York: Plenum.

Ward, G.C. (1985). *Before the trumpet: Young Franklin Delano Roosevelt, 1882–1905*. New York: Author.

Ward, G.C. (1989). *A first-class temperament: The emergence of Franklin Delano Roosevelt*. New York: Author.

Warraq, I. (2003). *Why I am not a Muslim*. New York: Prometheus.

Watson, P. (2005). *Ideas: A history from fire to Freud*. London: Weidenfeld & Nicolson.

Watts J. (1995) .*God, Harlem U.S.A.: The Father Divine story*. Los Angeles: University of California Press.

Weber, M. (1946). The sociology of charismatic authority. In H. Gerth & C. Wright Mills (Eds. and Trans.), *From Max Weber: Essays in sociology*. New York: Oxford University Press.

Weber, M. (1958). The three types of legitimate rule. *Berkeley Publications in Society and Institutions, 4*, 6–15.

Weber, M. (1964/1947). *The theory of social and economic organization* (A.M. Henderson & Talcott Parsons, Trans.). New York: Oxford University Press.

Weber, M. (1968a). *Economy and society* (Vols. 1–3). Edited by G. Roth & C. Wittich. New York: Bedminster Press.

Weber, M. (1968b). *On charisma and institution building*. Edited by S.N. Eisenstadt. Chicago: University of Chicago Press.

Webster, R. (1996). *Why Freud was wrong: Sin, science and psychoanalysis*. London: Harper Collins.

Weiner, D.N. (1998). *Albert Ellis: Passionate skeptic*. New York: Praeger.

Weir, A. (1999). *The life of Elizabeth I*. New York: Ballantine.

Wheelis, A. (1999). *The listener: A psychoanalyst examines his life*. New York: Norton.

Wilner, A.R. (1968). *Charismatic political leadership: A theory*. (Research Monograph No. 32). Princeton, NJ: Center for International Studies, Princeton University.

Wilner, A.R. (1984). *The spellbinders: Charismatic political leadership*. New Haven, CT: Yale University Press.

Wilson, A. (2008, August 19). Germaine Greer's essay is racist: Marcia Langton. *The Australian*, p. 5.

Wolpert, S. (2001). *Gandhi's passion: The life and legacy of Mahatma Gandhi*. New York: Oxford.

Worsley, P. (1970). *The trumpet shall sound*. St. Albans, UK: Paladin.

Wyden, P. (1988). *The unknown Iacocca.* London: Sidgewick and Jackson.

Yalom, I. (2002). *The gift of therapy.* New York: Harper Perennial.

Yeats, W.B. (1996). The second coming. In *The collected poems of W.B. Yeats.* New York: Scribner.

Zablocki, B. (1980). *Alienation and charisma: A study of contemporary American communes.* New York: Free Press.

Zerzan, J. (1988). *Elements of refusal.* Seattle: Left Bank Books.

Zimbardo, P. (1970). The human choice: Individuation, reason and order versus deindividuation, impulse and chaos. In W. Arnold & M. Levine (Eds.), *Nebraska Symposium on Motivation, 1969.* Lincoln: University of Nebraska Press.

Zohar. D. (2002). The effects of leadership dimensions, safety climate, and assigned priorities on minor injuries in work groups. *Journal of Organisational Behaviour, 23,* 75–92.

• • • • •

INDEX

• • • • •

www.ingramcontent.com/pod-product-compliance
Lightning Source LLC
Chambersburg PA
CBHW070544270326
41926CB00013B/2197